Index of Drinking Water Adequacy (IDWA)

International and Intra-national Explorations

Index of Drinking Water Adequacy (IDWA)
International and Intra-national Explorations

Edited by

Seetharam Kallidaikurichi E. and Bhanoji Rao

RIDGE BOOKS

*Published by NUS Press under the Ridge Books imprint for the
Institute of Water Policy, Lee Kuan Yew School of Public Policy,
National University of Singapore*

© 2010 The Institute of Water Policy

Ridge Books
an imprint of NUS Press
National University of Singapore
AS3-01-02, 3 Arts Link
Singapore 117569

Fax: (65) 6774-0652
E-mail: nusbooks@nus.edu.sg
Website: http://www.nus.edu.sg/nuspress

ISBN 978-9971-69-531-6 (Case)
       978-9971-69-530-9 (Paper)

Cover: The map indicates the lower the IDWA, relatively more serious is the problem of drinking water adequacy. While Developing Asia may seem catching up, there are large areas in Africa requiring attention.

Typeset by: Scientifik Graphics
Printed by : Mainland Press Pte Ltd

Contents

Preface

This monograph is the result of the collective efforts of four full time researchers of the Institute of Water Policy (IWP). It is motivated by the fact that water is central to economic and human development and its availability and access ought to be measured and monitored — an issue that has become even more crucial in the context of the Millennium Development Goals.

IWP research is conducted around a well thought out Medium Term Research Program. One component of the program deals with indicators and statistics, with the present monograph reporting on some of the research results.

IDWA first appeared in the *Asian Water Development Outlook, 2007* issued by the Asian Development Bank. Extensions and extrapolations have been attempted since and the key results are reported in this monograph.

IWP and the authors of this volume present IDWA to the policy makers and multilateral agencies in the hope that they find the idea useful and will help fine-tune it further. On its part IWP plans to annually update the IDWA for as many nations of the world as possible and present it as a handy flier and also upload the index on its website. The updated global IDWA flier will be available each year in June at the time of the Singapore International Water Week.

Two of the papers in this volume, the one dealing with Asian economies and the other dealing with Indian states, have appeared in *Water Policy,* 2010 and *Journal of Infrastructure Development,* 2009 respectively. We are grateful to the publishers of the journals for permission to reproduce the papers with minimal modifications.

We would also seek the understanding of the readers for some inevitable repetitions in the papers that are preserved simply to ensure the continuity of the thought process in each paper.

Seetharam Kallidaikurichi E. & Bhanoji Rao
April 2010

1

Introduction and Motivation

Bhanoji Rao and Fan Mingxuan

The Index of Drinking Water Adequacy (IDWA), as its name implies, is a monitoring tool devised to assess and benchmark countries' performance in providing *adequate drinking water* to their citizens. IDWA is inspired by the lack of an adequate indicator for monitoring access to drinking water of adequate quantity and quality. Rest of this chapter will focus on: (a) limitations of globally monitored indicators; (b) Water Poverty Index (WPI) and its limitations; and finally (c) the thinking behind IDWA.

Limitations of the Indicator Used for MDG Monitoring

Globally, the water component of Target 10 of the Millennium Development Goals (MDG) is measured as *proportion of population with sustainable access to safe drinking water*. WHO and UNICEF provide monitoring of progress on this target, and their definition of "sustainable access to safe drinking water" is given in Box 1.

Box 1

What does sustainable access to safe drinking water mean?

The Joint Monitoring Program (JMP) of WHO and UNICEF defines safe drinking water as follows:

• *Drinking water* is water used for domestic purposes, drinking, cooking and personal hygiene;

• *Access* to drinking water means that the source is less than 1 kilometre away from its place of use and that it is possible to reliably obtain at least 20 litres per member of a household per day;

• *Safe drinking water* is water with microbial, chemical and physical characteristics that meet WHO guidelines or national standards on drinking water quality;

• *Access to safe drinking water* is the proportion of people using improved drinking water sources: household connection; public standpipe; borehole; protected dug well; protected spring; rainwater.

Source: WHO (2004). Available at <http://www.who.int/water_sanitation_health/mdg1/en/index.html>

Although the definition is sophisticated enough to cover four different dimensions of "access", "proportion of people using improved drinking water source" is the only indicator applied in the actual monitoring process. The United Nations Development Group (UNDG, 2003) explained that "the indicator monitors access to improved water sources based on the assumption that improved sources are more likely to provide safe water". UNDG further clarified that "access and volume of drinking water are difficult to measure and so sources of drinking water that are thought to provide safe water are used as a proxy". Can the aforementioned argument justify the use of a single and simplistic indicator?

Figure 1 shows how a group of tribal women draw water from an agricultural well at Govind Tanda in Karepalli Mandal of Khammam

District, Andhra Pradesh, India. Despite the unknown quality and quantity of water they accessed and the difficulty in physically accessing water, this group of people might be considered as having access to improved water source due to the simple fact that they are using water from a well.

Figure 2 is the picture of the Hyderabad metro, at a place known as Puranapul in the city; tankers provide the water since the public taps had gone dry in early March 2010. Although WHO made it clear that water

Figure 1

Source: The Hindu, March 10, 2010.

vendors and tankers are not considered as improved drinking water source, these people might still be counted as with access to improved

Figure 2

Source: The Hindu, March 10, 2010.

Figure 3: July 28, 2009, in Chifeng City of Inner Mongolia, residents are collecting water from water wagons because the tap water is contaminated by an overflow of rainwater. The tap water was not cut off, but the government has warned residents not to drink tap water.

Source: Xinhua News, July 30, 2009.

drinking water because they have public taps, no matter the taps work or not.

Similar stories of what may be considered as inadequate access are very common across the developing world (see Figures 3 and 4), and if not addressed properly, will seriously undermine the achievement of MDG within the set time frame.

Quest for a Comprehensive Indicator: The WPI

The drinking water MDG being spelt out in an extremely simple way, it cannot and is not meant to indicate significant and sustained progress on the drinking water front. MDG for water as it stands now is about access [almost any access]. There is no concern for sustained access; no concern for quantity and quality; there is no linkage between access and resource availability; etc. These and allied considerations have motivated the design and estimation

Figure 4: March 29, 2007, Jinxiang County of Sichuan Province. Villagers collect water from a well. Due to drought, each villager can get only half a barrel of water per day.

Source: *Xinhua News*, March 29, 2007.

of the Water Poverty Index (WPI), briefly explained below and evaluated in the next section. The pioneering work on WPI design and implementation can be accessed from Lawrence, Meigh and Sullivan (2002), Sullivan and Meigh (2003a, 2003b) and Sullivan, Meigh *et al.* (2003c). The following extracts from the last cited paper from *Natural Resources Forum*, a journal of the United Nations brings out the motivation for developing WPI.

> Increasingly, water is seen as one of the most critically stressed resources, and much attention is now being paid to global water stress and the water needs of the poorest.... [Sullivan, Meigh *et al.* (2003c), p. 189]
>
> In this context, that is, in order to assist with the global effort to tackle water problems ... monitoring tools are required.... Currently, monitoring access to water and sanitation is carried out at the international level by the WHO and UNICEF joint

monitoring program for water supply and sanitation.... These data provide much valuable information, however they include many simplifications and have a number of shortcomings.... In order to move towards a truer and more comprehensive assessment of the situation, we suggest that a monitoring tool is needed that looks at water availability and access in a broader way. [Ibid., p. 190]

Water Poverty Index: An Evaluation

WPI combines measures of water availability and access with measures of people's capacity to access water on a sustained basis, the use of water, and environmental factors that affect water quality and ecology. The five *major* components of WPI are briefly explained below and a bit more detail on each is given thereafter.

- *Availability* of water is indicated by surface and groundwater resources that can be drawn upon by the residents of a nation.
- *Access* includes both safe water for drinking and cooking, and water for irrigating crops or for nonagricultural use.
- *Capacity* is taken as the possession of purchasing power to obtain improved water. Additionally, in the construction of WPI, education and health are considered because they "interact with income and indicate a capacity to lobby for and manage a water supply". (Lawrence *et al.* 2002, p. 2)
- *Use* reflects domestic, agricultural, and nonagricultural uses.
- *Environmental factors* considered are those likely to impact on regulation and affect capacity.

For *resource availability* both external and internal inflows (resources) are considered. The volumes are measured on a per capita basis and converted to a log scale.[1] Weights for the two sets of inflows are half and half. While it is undoubtedly true that resources

---

[1] The justification given for taking the log scale was "to reduce the distortion caused by high values".

should not be just limited to internal resources, we believe that the data on external resources could be flawed and hence may not do justice when included as a component in a composite index such as WPI. To substantiate the difficulty in obtaining good data on external inflows, we quote the World Bank's World Development Indicators (WDI), 2006 (p. 148): "River flows from other countries are not included because of data availability." The inflows data in WDI, especially those that include use-wise distribution of withdrawals, were all given for a range of years, namely 1987–2002. Thus, if the World Bank thought that data were not available even within this broad time range, the matter must be taken seriously and rough and ready data should be eschewed from serious estimates. In our view, therefore, it is best to limit the measurement of availability to internal resources, notwithstanding the fact that some countries may have substantial external inflows.

In the computation of WPI, *access* has three components: percentage of population with access to 'safe' water, percentage of population with access to sanitation, and an index that relates irrigated land as a proportion of arable land to internal water resources. "This is calculated by taking the percentage of irrigated land relative to the internal water resource index and then calculating the index of the result. The idea behind this method of calculation is that countries with a high proportion of irrigated land relative to low internal available water resources are rated more highly than countries with a high proportion of irrigated land relatively to high available internal water resources." (World Bank's WDI 2006, p. 6)

Our comments on the access indicators are as follows. It is important to look at access for water and not mix it up with sanitation, especially because only one sanitation indicator is available.[2] As for water availability for irrigation, it is important, but a relatively more direct indicator could be considered. Water access indicators

---

[2] One must not give the impression that taking just one crude indicator on sanitation is good enough for taking care of the vital human need for sanitation.

and the percentage of internal resources used for irrigation could be considered for measuring access, if access is to take account of water for personal and agricultural uses. In general, water for personal use (drinking water) should be distinguished from other uses. In particular, water used for agriculture may be better suited for composite indicators, such as an index of food security. One could safely say, at least at the present state of knowledge that drinking water has no substitute. In sharp contrast, food can be imported from those nations that are water and land abundant by those that have little of them. It does not make sense to mix up water for personal and agricultural uses.

The ***capacity measure*** in WPI has four components. The most straightforward and easily justifiable component is the index of purchasing power denoted by log of gross domestic product (GDP) per capita in purchasing power parity (PPP) terms. The other three indicators are under-5 mortality rate (per 1,000 live births), an index of education taken from the Human Development Report 2001, and the Gini coefficient, which denotes the extent of inequality in income/expenditure. Where the Gini is not available, measurement is based on the first three components. As is well known, HDI captures precisely the indicators of income, life expectancy, and education. Thus, one can use HDI itself as a measure of capacity, if one is keen to add the three components.

Our judgment is that it is enough to use income as the capacity indicator. As for education, the link between education and capacity is circumspect. Regarding health, water — the elixir of life — is an input and health could be considered as output. Thus, the measurement of capacity could be simply limited to income. As for Gini, use of household income in some cases and household expenditure in others, variations in household size across income and expenditure groups, differences in the years to which different household surveys refer, accuracy of income/expenditure measurement, etc., severely limit the comparability of the indicator across countries.

We conclude that capacity to buy water is well reflected by per capita income. That alone is good enough as the case of Singapore illustrates. This country has the buying power to obtain some water from neighboring Malaysia. In addition, Singapore's buying power has been responsible for experiments with obtaining water from desalination and recycling.[3]

In WPI *use* is measured on the basis of three components: domestic water use per capita (cubic meters per person per year), industrial water use per capita, and agricultural water use per capita. Taking 50 litres per person per day as a reasonable target for developing countries, an index is developed taking countries below and above the norm. The typical nation with consumption of 50 litres has an index value of 1. Countries below the norm have an index that is proportionately reduced. For countries above the minimum, the index decreases as consumption exceeds 50 by higher margins. This procedure is used in order not to 'reward' excessive use of water. Use, however, is not limited to personal use, but also encompasses industrial and agricultural uses (in addition to irrigation water noted earlier).

For industrial and agricultural water use, instead of taking per capita consumption, the proportion of GDP generated by the sector is divided by the proportion of water used by the sector. The authors of WPI state, quite rightly, that the index for each sector reflects the efficiency of water use. It is efficiency that is measured this time around and not poverty!

If the intention is to measure water poverty or inadequacy for personal use, there is little justification to include other uses, which should go into other more appropriate composite indicators. Indicators of agricultural sustainability and efficiency, for instance, could consider the extents of water available and used in agriculture. Similarly since industrial location decisions are based on various factors and these include water availability, indicators of efficiency of water use (and reuse etc.) are of great importance. These interesting additional components do help promote WPI that "looks at water

---

[3] For a comprehensive description and analysis of the Singapore case, see Tortajada (2006).

availability and access in a broader way". Our opinion, however, is that when a composite index becomes too broad and too much of a mix of disparate components, it then loses capability to convey the right messages and be a sound tool for policy formulation and progress monitoring purposes.

The index of **environment** in WPI is an average of 5 composite components, all of which are based on data used for the Environmental Sustainability Index (ESI).[4] The five are: an index of water quality (based on dissolved oxygen concentration, phosphorus concentration, suspended solids, and electrical conductivity); an index of water stress (based on fertilizer consumption per hectare of arable land, pesticide use per hectare of crop land, industrial organic pollutants per available freshwater, and the percentage of country's territory under severe water stress); index of regulation and management capacity (based on indices of environmental regulatory stringency, environmental regulatory innovation, percentage of land area under protected status, and the number of environmental impact assessment guidelines for different sectors of the economy); index of informational capacity (based on measures of availability of sustainable development information at the national level, environmental strategies and action plans, and the percentage of ESI variables missing from public global data sets); and an index of biodiversity (based on the percentage of threatened mammals and birds).

Water poverty/inadequacy is intimately linked to water quality and there can be no two views on that. Other environmental measures are all important for ESI but perhaps not water for personal use per se. Water quality indicators alone should be considered for measurement of water poverty. After all, managerial and system efficiency, etc., are ultimately to provide adequate and good quality water. Quantum indices having been covered earlier, an index of quality should also be included. Yet, in regard to Asia, the poor extent of

---

[4] World Economic Forum *et al.* (2001), quoted in the references cited in footnote 5.

data availability (for just about a dozen economies) precludes the use of the water quality indicators.[5]

Box 2 gives a *summary of components and subcomponents* of the WPI, with the number of indicators in each component indicated in square brackets. There are 12 indicators in all for resources, access, capacity and use, in sharp contrast to the 15 for environment. It is not proper to say which of the indicators are more or less important unless the purpose and focus are clearly marked.

What does an average of 27 indicators denote? We leave it to the reader to explore. One must yet be grateful to the designers of WPI for bringing into focus the need to go beyond the simple access indicators for water for personal use. There is considerable scope to refine the inputs of WPI and redesign the output in the form of a composite index that is relatively more closely linked to access to drinking water, one of the most vital components in delivery of inclusive growth. Also, in the spirit of HDI, it is necessary to limit the number of components in a way that would tell the policy maker why a nation's composite index is relatively low and what component needs priority attention.

---

[5] The following tabulation is extracted from the ESI database readily available on the internet.

Country	Dissolved Oxygen Concentration	Electrical Conductivity	Phosphorus Concentration	Suspended Solids
Bangladesh	6.7	231.6		4.08
Cambodia		13.62	0.04	4.03
PRC	8.62	522.78	0.28	7.97
India	6.43	2,240.7	0.2	1.83
Indonesia	3.31	167.13	0.57	5.37
Lao PDR	6.96	20.88	0.12	4.4
Pakistan	6.77	492.46	0.67	5.54
Philippines	7.42	136.7		3.81
Republic of Korea	11.01	145.29	0.13	2.21
Taipei, China	6.1	2,244	0.18	5.25
Viet Nam	5.3	559.87	0.12	4.63

Box 2

The 27 Indicators in the Water Poverty Index

Resources: Per capita external and internal inflows [1]

Access: Percentage of population with access to 'safe' water, percentage of population with access to sanitation, and an index that relates irrigated land as a proportion of arable land to internal water resources [3]

Capacity: An index of purchasing power, under-5 mortality rate, an index of education, and the Gini coefficient [4]

Use: Domestic water use per capita, industrial water use per capita, and agricultural water use per capita [3]

Environment: Water quality (four indicators), water stress (four), regulation and management (four), informational capacity (three), and biodiversity (one) [16]

Index of Drinking Water Adequacy (IDWA)

To focus attention on the critical area of drinking water and to reduce the data requirements to a reasonable level, we developed IDWA which is based on a handful of essential indicators: resource availability, access to improved drinking water sources, capacity to buy water, water quality and water usage. This monograph introduces methodology used in IDWA and its application to international, regional and national levels.

Rest of the monograph, following this short paper on the thinking behind IDWA, is organized as follows. The second paper provides the index for 23 Asian economies, while the third is about IDWA-Global (IDWA-G) for 144 economies of the world, which from now on will appear annually on the Institute of Water Policy website. Intra-country studies for China, Indian and Vietnam are in papers 4, 5 and 6 respectively. Concluding notes (by the editors) are put out at the end.

References

Lawrence, P., Meigh, J.R., and Sullivan, C.A. (2002). The Water Poverty Index: an International Comparison. *Keele Economics Research Papers, 2002/19*.

Sullivan, C.A. and Meigh, J.R. (2003a). Considering the Water Poverty Index in the context of poverty alleviation, *Water Policy*, 5, 513–28.

———— (2003b). Access to water as a dimension of poverty: the need to develop a Water Poverty Index as a tool for poverty reduction. In: Olcay Ünver I.H., Gupta R.K. and Kibaroðlu, A. (eds.), *Water Development and Poverty Reduction*, Kluwer, Boston, 31–52.

Sullivan, C.A., Meigh, J.R., *et al.* (2003c). The Water Poverty Index: Development and application at the community scale, *Natural Resources Forum*, 27, 189–99.

Tortajada, C. (2006). Water Management in Singapore, *Water Resources Development*, 22 (2), June, 227–40.

United Nations Development Group (2003). *Indicators for monitoring the Millennium Development Goals: Definitions, Rationale, Concepts and sources.* New York: United Nations.

World Bank (2006). *World Development Indicators.* Washington: World Bank.

World Economic Forum, Yale Center for Environmental Law and Policy and Center for International Earth Science Information Network (2001). *2001 World Environmental Sustainability Index.* Davos, Switzerland.

World Health Organization (2004). *Health through safe drinking water and basic sanitation.* Available at <http://www.who.int/water_sanitation_health/mdg1/en/index.html> [Accessed March 16, 2010].

2

Drinking Water Adequacy Differentials in Asia[1]

Seetharam Kallidaikurichi E. and Bhanoji Rao

An Index of Drinking Water Adequacy (referred to as IDWA-I in this paper) was first proposed in 2007 for 23 member countries of the Asian Development Bank (ADB), and formed part of the Asian Water and Development Outlook (AWDO), 2007 brought out by the ADB. IDWA-I was obtained by averaging 5 separate component indicators referring to capacity to buy water, extent of resource availability, amount of water used, water quality (indicated by a proxy variable, namely the death rate due to diarrhea) and the per cent of people with access. This paper reports the main results of IDWA-I and IDWA-II, in which we replace general

[1] An earlier (lengthier) version of this paper is available as a Discussion Paper along with the *Asian Water and Development Outlook* (AWDO) *2007*, issued by the Asian Development Bank. The authors acknowledge the encouragement and constructive comments received from the intellectual leader of AWDO, Professor Asit Biswas, President and Academician, Third World Center for Water Management, and Distinguished Visiting Professor at the Lee Kuan Yew School of Public Policy, National University of Singapore. The usual disclaimers apply. The authors also acknowledge the kind comments of an anonymous referee, which have significantly added value to the paper.

access with specific access via home connection, after finding out the relatively weak correlation between the two types of access. Because of the dominating influence of the other common components, IDWA-I and IDWA-II are highly correlated indicators. The two, however, bring out diverse relative ranks of different countries.

1. Introduction

As Paul Streeten (1994, p. 295) has put it succinctly: "... indices are useful in focusing attention.... They have considerable political appeal.... They are eye-catching." This paper reports on the concept, method and results on an indicator of drinking water adequacy, and provides the indicator values and implications for 23 Asian economies at the turn of this century.

Presently the most widely used indicators relating to drinking water refer to the extent of access denoted by the proportion (percentage) of total population with access to safe drinking water. Thus, even the well publicized and relatively recent Millennium Development Goal on water and sanitation (MDG-7)[2] hopes to "halve by 2015 the proportion of people without sustainable access to safe drinking water and basic sanitation".

If the access percentages referred to earlier are too simplistic, there is one other that is most complex, an index known as the Water Poverty Index that is based on 27 separate indicators. The concept and methodology of WPI has been amply explained in

---

[2] The 8 MDGs, endorsed by 191 nations in all, are Eradicate Poverty & Hunger; Achieve Universal Primary Education; Promote Gender Equality; Reduce Child Mortality; Improve Maternal Health; Combat HIV AIDS, Malaria and Other Diseases; Ensure Environmental Sustainability; and Develop a Global Partnership for Development. The sub-goals of Goal 7 are: Integrate the principles of sustainable development into country policies and programs; Reverse loss of environmental resources; Reduce by half the proportion of people without sustainable access to safe drinking water; and Achieve significant improvement in lives of at least 100 million slum dwellers, by 2020.

Lawrence, Meigh and Sullivan (2002).[3] It has been pointed out that the idea behind WPI is to combine measures of water availability and access with measures of people's capacity to access water on a sustained basis, the use of water and environmental factors, with an impact on water quality and ecology.

The idea behind WPI is good and laudable, but the data requirement (27 indicators) is mind-boggling. Accepting the basic idea of the utility of a composite index with key parameters going into it, and limiting ourselves to the most appropriate indicators, the Index of Drinking Water Adequacy (referred to as IDWA-I from now on) was proposed originally for use in the maiden issue of the *Asian Water Development Outlook*, 2007 (A.K. Biswas and K.E. Seetharam, 2008).

The objectives of this paper are essentially twofold: to place IDWA-I before a wider audience and to report results on IDWA-II. The paper is organized as follows. The making of IDWA-I is explained in the next section and inter-country variations in the index are briefly touched upon. In Section 3, IDWA-II, an improvement over the earlier version, is attempted and the inter-country variations are analyzed. In Section 4, after a brief review of presently available and globally publicized data on access to water and sanitation, we appeal to the international development community to invest time and resources towards improving data quality.

2.   Index of Drinking Water Adequacy (IDWA-I)

The index ought to reflect the key variables determining the availability of adequate amount of drinking water of the right quality. First, ensuring adequate drinking water is facilitated if the nation has its own resources. Second, one or the other types of access mechanisms will help in ensuring supply. Third, adequacy is best

---

[3] There have been a few further explorations on WPI by the authors and these are Sullivan, Meigh and Lawrence (2005), Sullivan and Meigh (2003a, 2003b), and Sullivan (2003).

assured with a relatively higher, than lower, income level, that is, the capacity to purchase water. Fourth, despite having resources, access mechanisms and capacity, actual use of water may be adequate or less than adequate. Finally, what is sought is not just any drinking water, but water of good quality.

Based on the aforementioned rationale, it is surmised that the key components of IDWA should reflect resources, access, capacity, use and quality, explained below (see also Annexes).

Resources

Estimates of renewable internal fresh water resources[4] per capita were taken from the World Development Indicators (WDI) 2006, which refer to 2004. The per capita figures are converted to a log scale. The resulting values are converted to an index as follows. Taking the resource per capita, Rj for country j, we have

$$\text{Resource Indicator for country 'j'} = \left[\frac{(\log R_j - \log R_{min})}{(\log R_{max} - \log R_{min})} \right] \times 100 \quad [1]$$

The maximum recorded per capita internal resource is that of Papua New Guinea (PNG) at 138,775 cubic meters (m^3) in 2004. The PNG figure is the maximum not only for Asia but also for all countries covered in WDI. The minimum possible resource per head is taken at a nominal 1 m^3, which in log form is zero.

Access

This refers to access as under the MDG.[5] We used the 2004 estimates of access, measured as percent of population with access to

---

[4] Internal Renewable Water Resources (IRWR) comprises the average annual flow of rivers and recharge of groundwater (aquifers) generated from endogenous (internal) precipitation. Natural incoming flows originating outside a country's borders are not included. The total flow is given in billion cubic meters and per capita flow in cubic meters.

[5] **Definition:** MDG refers to "Percentage of the population using improved drinking water sources (including household water connection, public standpipe,

a sustainable 'improved' water source. In this case, the maximum possible access is availability of safe water for 100% of the population. The minimum is not taken as zero, since it is a demographic-geographic impossibility, given that people need at least some minimum drinking water in any settlement. In the MDG database, among 225 economies, Ethiopia had the lowest access rate of 22% in 2004, while countries with 100% access ranged from Andorra to Malaysia[6] and USA.

$$\text{Access Indicator for country 'j'} = \left[\frac{(A_j - 22)}{(100 - 22)} \right] \times 100 \qquad [2]$$

Capacity to Buy Water

Per capita GDP in PPP dollars is used as a measure of a nation's capacity (C) to produce/purchase and supply adequate amounts of drinking water. Among the 23 economies, estimates are not available for Myanmar and Turkmenistan. In both cases, estimates are derived by comparing a nearby economy on the scale of power consumption per head. Thus, comparing Bangladesh and Myanmar, the per capita GDP of the latter is obtained. In the case of Turkmenistan, the comparison was with Uzbekistan. As for the minimum-maximum estimates, the figure of US$630 of Malawi was used as

---

borehole, protected dug well, protected spring, rainwater collection and bottled water — if a secondary available source is also improved). **Method of Computation:** Data from household surveys and censuses are adjusted to improve comparability over time. Survey and census data are then plotted on a time scale from 1980 to present. A linear trend line, based on the least-squares method, is drawn through these data points to estimate coverage for 1990 and latest available year." *Source:* <http://mdgs.un.org/unsd/mdg/Metadata.aspx?IndicatorId=0&SeriesId=667>.

[6] The Malaysian rate was shown as 99 in the UN database. It is rounded off to 100 here.

the minimum and US$20,530 of the Republic of Korea was the maximum.[7]

$$\text{Capacity Indicator for country 'j'} = \left[\frac{(\log C_j - \log C_{min})}{(\log C_{max} - \log C_{min})} \right] \times 100 \quad [3]$$

Use

The most challenging task has been the computation of per capita water consumption by the domestic sector, which, in this paper, is referred to as 'drinking water', and for which a set of readymade numbers is unavailable from any of the international databases. Thus, three alternative sets were computed and averaged for the final estimate.

The first set is based on the WDI 2006 data for each country on annual freshwater withdrawal[8] in billion m³ for '1987–2002', which refers to some year within the range, depending on data availability. Dividing the aggregate by the average population for 1990 and 2000 and multiplying the per capita withdrawal by the proportion used for domestic purposes (estimates for which are also available in WDI), we obtain the annual use per capita. The annual figure is converted into litres per capita per day (LPCD).

The second and third sets of estimates are based on the data reported by World Resources Institute[9] for some year in the range 1987–95, and for the year 2000, respectively. In both cases, the starting point is the annual withdrawal per capita. Applying the percentage of water used for domestic purposes and converting the

---

[7] We could have used the global maximum, the US figure of $40,000. However, our aim, as far as possible, is for one of the DMCs to have the maximum index value of 100, except when unjustified. In the case of per capita income, as long as the level is high enough, it reflects the capacity to procure water, even if internal water resources are not available.

[8] *Annual Total Water Withdrawals* is the gross amount of water extracted annually from any source, either permanently or temporarily, for a given use. It can be either diverted towards distribution networks or directly used. It includes consumptive use and conveyance.

[9] The authors are grateful to Narciso Prudente (ADB Research Assistant for the AWDO project) for the data compilation.

annual figure to daily average, estimates of consumption per capita per day are obtained. Estimates for 2000 were available only for 10 countries. Thus, the final average per capita consumption (U) is based on 3 observations for 10 economies and 2 observations for the remaining 13 economies.

For obtaining the index of use based on the estimate of per capita consumption, we require minimum and maximum norms. The minimum is taken as 70 LPCD as prescribed by the Indian Government.[10] For the maximum, although there are countries within our sample that have recorded consumption levels as high as 393 (Republic of Korea), we base our norm on the laudable experience of Singapore, where water conservation is combined with guaranteed continuous supply of water that can be safely consumed straight from the tap. The 1995–2002 average of per capita domestic consumption in Singapore was found to be 167 LPCD. Using the minimum and maximum norms, the index of use was computed for the 23 economies.

$$\text{Use Indicator for country 'j'} = \left[\frac{(U_j - 70)}{(167 - 70)} \right] \times 100 \qquad [4]$$

The computed index for use is higher than 100 in some cases due to relatively high levels of average annual per capita consumption. In such cases the value is taken as 100, thus ignoring what may well be some 'wasteful' consumption (in comparison to Singapore, of course) due to lack of appropriate conservation measures, which could include progressive pricing policies. There are cases where the index is negative, simply because the numerator is negative, that is, the particular country's per capita consumption is less than the norm of 70 LPCD. These negative index values are left untouched, because they speak eloquently about water inadequacy.

---

[10] Under the Indian Government's Accelerated Urban Water Supply Programme taken up in the Eighth Five Year Plan (1992–97), the norm for drinking water was set at 70 LPCD. Other norms were given as follows: with sewerage: 125 LPCD; without sewerage: 70 LPCD; and with spot sources and public stand posts: 40 LPCD.

Quality

In the absence of reliable national data on water quality, we use data on diarrheal deaths, as a proxy. Data on diarrheal death rate (DR) expressed as diarrheal deaths per 100,000 people for the year 2000 are considered. World Health Organization data indicate a maximum DR close to 100 in Lao PDR and a minimum of 0.5 in the Republic of Korea, with a wider band of variation in the world at large.[11] The index is computed by taking the difference between 100 and the country value, as an indirect measure of water quality.

$$\text{Quality Indicator for country 'j'} = \left[\frac{(100 - DR_j)}{(100 - 0)} \right] \times 100 \qquad [5]$$

$$= 100 - DR_j$$

3. IDWA-I Values for 23 Economies

Excluded and Included Economies

A total of 44 Developing Member 'Countries' (economies) or DMCs are part of the membership of the Asian Development Bank. Not all the members could be covered in this exercise due to paucity of adequate information from national and global sources.[12] Though the absolute number of economies included was just a little over half of the total number of DMCs, the 23 covered DMCs account for 3.4 billion people (2004 estimate), close to 99% of the total population of all 44 economies. For each of the 23 economies, the Resource, Access, Capacity, Use, and Quality indicators are averaged to obtain IDWA values, presented in Table 1 in descending order.

---

[11] Among over 200 countries, the maximum of 370 was observed for Angola and 8 countries have estimates close to zero: Ireland, Lithuania, Portugal, Slovakia, Poland, Italy, Austria, and Czech Republic have each a level of 0.1.

[12] In all, 21 economies could not be covered, and these included Afghanistan; Armenia; Bhutan; Brunei Darussalam; Hong Kong, China; Singapore; Taipei, China; and several of the island and other economies with less than a million people. [See, however, the write up on Singapore that follows.]

Addressing Possible Reservations on IDWA

The index provides the relative position of the different DMCs in a more comprehensive fashion than simple access indicators. It is likely, however, that researchers as well as policy makers might have strong reservations on the relative rankings and the included components. For instance, one could legitimately point out that the absolute value of IDWA for Malaysia should be close to 100, since the most critical components, namely, Access, Use and Quality indicators are close to 100, and that the country's IDWA has been brought down artificially by including Resources and Capacity. That this is not a valid criticism is illustrated by the case of Singapore noted below.

Table 1: IDWA-I for 23 Asian Economies

	Resource	*Access*	*Capacity*	*Use*	*Quality*	*IDWA*
Malaysia	85	100	79	100	99	92
Korea, Rep. of	61	90	100	100	99	90
Philippines	73	81	59	100	84	80
Viet Nam	71	81	42	100	87	76
Kazakhstan	72	82	69	50	98	74
Azerbaijan	58	71	52	100	89	74
Kyrgyz Republic	77	71	31	100	87	73
Thailand	68	99	73	32	93	73
Uzbekistan	54	77	31	100	98	72
Turkmenistan	48	64	31	93	77	63
Tajikistan	78	47	18	100	67	62
PRC	65	71	64	16	92	61
India	60	82	46	56	57	60
Indonesia	80	71	49	13	84	59
Mongolia	81	51	34	31	66	53
Sri Lanka	66	73	55	−35	96	51
Pakistan	49	88	36	0	21	39
Nepal	75	87	25	−31	32	38
Bangladesh	56	67	33	−22	53	37
Myanmar	83	72	26	−52	49	35
Papua New Guinea	100	22	37	−58	57	32
Lao PDR	88	37	31	−7	2	30
Cambodia	77	24	37	−56	14	19

Note: Background tables are given in the Statistical Appendix.

IDWA for Singapore

The country scores 100 on each of 4 IDWA components (access, capacity, use, and quality). Yet, because of an index value of 42 for the resource component,[13] IDWA based on 5 components is 88, lower than the values for Malaysia and Republic of Korea (92 and 90, respectively). The numbers illustrate rather clearly how important it is to include not only a measure of resources but also capacity.[14] In the case of Singapore, though resources are scarce, its capacity reflected in the high level of per capita GDP, has enabled the nation to innovate constantly and be in the forefront of nations that have achieved 24×7 supply of top quality drinking water, consumed directly from the tap at home.

Indeed, the merit of IDWA is its 'eye catching' value and revealing composition. Thus, PNG has the resource edge, but not the rest of the wherewithal to supply adequate drinking water to its population. Malaysia has an edge over the Republic of Korea, because of the former's high degree of resources and access, while the latter has a high level of capacity that soon must be converted into higher access. Similarly, a comparative assessment of IDWA for the two most populous nations of Asia and the world, namely, PRC and India is also illumining. Despite almost identical values for IDWA, there are stark differences in some of the components. The comparatively high use index in India, for instance, does not mean much when one takes cognizance of relatively poor water quality.

---

[13] Singapore's internal water resources are placed at 142 cubic meters per capita in 2004 as per the World Development Indicators, 2006. Relative to PNG (value: 138775), based on the respective logarithmic values, Singapore resource index works out to 42. [Log of 138775 = 5.142, log of 142 = 2.152, and the index is .4185 × 100 = 41.85 or 42 rounded.]

[14] If Singapore were to produce all the food it consumes, it would have needed a lot more water than it could ever produce or buy. Thanks to its purchasing power, it enjoys the convenience of importing food, and indirectly the (virtual) water that goes into the food.

IDWA numbers are as good as the reliability of the data that go into them. The data inadequacy is most glaringly reflected in the fact that we have had to make do with diarrheal death rate as a proxy for water quality. The data gap here serves to underscore the urgent need for water testing on a routine basis in as many cities and villages as possible and ensuring the availability of the information.

4.   IDWA-II: Incorporating House Connections

IDWA-I was based on the premise that the MDG are set with reference to "access to a sustainable improved water source", even though such access would seem to be sub-optimal in terms of ensuring minimal health risks (Table 2). Indeed, if one were to factor in not only the health risk, but also the opportunity cost of time lost in collecting water and related influences, the intrinsic merits of water connection at home becomes indisputably clear. Despite this comment, it is not necessary to replace the access data with house connections data if cross-country and time-series trends in both look very similar, regardless of expected differences in the absolute values.

Table 2:   Optimal Access: House Connection

Service level	Distance/Time	Likely Volume of Water Collected	Health Risk
No Access	More than 1 kilometre, Over 30 minutes round trip	Very low, 5 litres per capita per day	Very high
Basic Access	Less than 1 kilometre and 30 minutes	About 20 litres per day	High
Intermediate Access	At least one tap in premises or close by	About 50 litres per capita per day	Low
Optimal Access	Water supply within house with more than one tap	100–200 litres per day	Very Low

Source: Adapted from WHO, 2004, *Domestic water quantity, service level and health*, Geneva: WHO (quoted in WHO-UNICEF, 2005).

To evaluate the issue, data on house connections are needed. The Water and Sanitation Information Website of the Joint Monitoring Program of the World Health Organization and the UN Children's Fund has country pages that provide the basic data available from surveys and censuses. Based on them, we could assemble, for one or more years between 1987 and 2003, data on the percentage of families served by house connections for water for 22 economies for urban areas and 20 for rural areas. Juxtaposing the data with respective averaged access rates for 1990 and 2004, the correlation coefficients obtained are 0.65 for urban and 0.40 for rural. The values are not high enough to justify the use of access as a proxy for house connections. This provided the motivation to work out IDWA-II.

In IDWA-II, we replace the single national access indicator with urban and rural house connection rates, thus raising the number of component indicators from 5 to 6, and focusing on the need to supply drinking water to urban and rural areas on the same footing if possible. To be sure, IDWA-I and IDWA-II are quite close, the linear correlation coefficient between them being a high 0.95. Given the dominance of other components (4 out of 6), this is not unexpected. However, on IDWA-II, the relative position of some of the economies has changed (Table 3). For instance, the second position on IDWA-II goes to Kazakhstan, instead of the Philippines, which has the second spot on IDWA-I. When the top and second spots are compared, the index value jumps down from 91 to 67 in regard to IDWA-II, while the IDWA-1 transition has been relatively smooth. This is also manifest in what could be the most interesting and significant difference between the two indexes: the extent to which they differ in each economy. As per the data in Table 4, there are relatively few economies that have a fairly close index value on IDWA-I and IDWA-II.

5. Concluding Observations

This paper illustrates a modest attempt to develop an indicator of drinking water adequacy. IDWA-I and IDWA-II portray the

Table 3: Economies Ranked by IDWA-II and IDWA-I

Economy	IDWA-II	Economy	IDWA-I
Malaysia	91	Malaysia	92
Kazakhstan	67	Philippines	80
Philippines	63	Viet Nam	76
PRC	61	Kazakhstan	74
Viet Nam	60	Thailand	73
Thailand	58	PRC	61
India	46	India	60
Indonesia	43	Indonesia	59
Mongolia	43	Mongolia	53
Sri Lanka	37	Sri Lanka	51
Papua New Guinea	33	Pakistan	39
Pakistan	29	Nepal	38
Lao PDR	27	Bangladesh	37
Nepal	26	Myanmar	35
Bangladesh	24	Papua New Guinea	32
Myanmar	21	Lao PDR	30
Cambodia	18	Cambodia	19

Note: The six component indicators of IDWA-II are in Table 8 of the Statistical Appendix.

Table 4: Percentage Excess of IDWA-I over IDWA-II

Economy	%	Economy	%
Myanmar	67	Thailand	26
Bangladesh	54	Mongolia	23
Nepal	46	Lao PDR	11
Sri Lanka	38	Kazakhstan	10
Indonesia	37	Cambodia	5
Pakistan	34	Malaysia	1
India	30	PRC	0
Philippines	27	Papua New Guinea	0
Viet Nam	27		

possibilities for the eventual development of good indicators for the householdwater sector, provided data of good quality[15] are forthcoming. Future research could encompass data auditing to figure out how to ensure timely data of good quality, extension of IDWA to all countries across the world, and fine-tuning the components and exploring IDWA's linkages to several key social and economic parameters that usually enter the discourse on development.

References

Biswas, A.K. and Seetharam, K.E. (2008). Achieving Water Security in Asia, *International Journal of Water Resources Development*, 2008.

Esty, D.C., Marc, L., Srebotnjak, T., and Alexander de Sherbinin (2005). *2005 Environmental Sustainability Index: Benchmarking National Environmental Stewardship.* New Haven: Yale Center for Environmental Law and Policy.

Lawrence, P., Meigh, J., and Sullivan, C. (2002). The Water Poverty Index: an International Comparison, Keele Economics Research Papers Number 2002/19, Department of Economics, Keele University, Keele, Staffordshire, UK.

Streeten, P. (1994). Human Development: Means and Ends, *American Economic Review*, 84, 2, 232–7.

Sullivan, C., Meigh, J., and Lawrence, P. (2005). Application of the Water Poverty Index at different scales — a cautionary tale, *Agriculture, ecosystems and environment.* [A search was made on the journal site but it has not become possible to find the article publication details.]

Sullivan, C.A. and Meigh, J.R. (2003b). Access to water as a dimension of poverty: the need to develop a Water Poverty Index as a tool for poverty reduction. In: Olcay Ünver I.H., Gupta R.K. and Kibaroðlu, A. (eds.), *Water Development and Poverty Reduction*, Kluwer, Boston, 3152.

_____ (2003a). Considering the Water Poverty Index in the context of poverty alleviation, *Water Policy*, 5, 513–28.

Sullivan, C.A. (2003). The Water Poverty Index: A new tool for prioritisation in water management. *World Finance.* [See parenthetical note under Sullivan, Meigh and Lawrence (2005).]

---

[15] The issue is addressed in paper 6 of this volume.

Tortajada, C. (2006). Water Management in Singapore, *Water Resources Development*, 22 (2), June, 227–40.

WaterAid (2004). *Manual for Valuing the Benefits of WaterAid's Water and Sanitation Projects*, Water Aid, July 2004.

WaterAid India (2005). *Drinking Water and Sanitation Status in India*. New Delhi: WaterAid India Country Programme.

Annexes

Annex 1: Resources

Country	Resource per capita	Log of Resource per capita	Resources Index
Azerbaijan	977	2.99	58.14
Bangladesh	754	2.88	55.95
Cambodia	8,738	3.94	76.65
PRC	2,170	3.34	64.88
India	1,167	3.07	59.64
Indonesia	13,043	4.12	80.03
Kazakhstan	5,030	3.70	71.98
Korea, Rep. of	1,349	3.13	60.87
Kyrgyz Republic	9,121	3.96	77.01
Lao PDR	32,878	4.52	87.84
Malaysia	23,298	4.37	84.93
Mongolia	13,839	4.14	80.53
Myanmar	17,611	4.25	82.57
Nepal	7,454	3.87	75.30
Pakistan	345	2.54	49.35
Papua New Guinea	138,775	5.14	100
Philippines	5,869	3.77	73.29
Sri Lanka	2,575	3.41	66.33
Tajikistan	10,311	4.01	78.04
Thailand	3,297	3.52	68.42
Turkmenistan	285	2.45	47.74
Uzbekistan	623	2.79	54.34
Viet Nam	4,461	3.65	70.97

Highest Value: 138,775 (Papua New Guinea)
Lowest: 1

Annex 2:   Access (Per cent of population with access to sustained improved water source), 2004

Country	Access	Access Index
Azerbaijan	77	71
Bangladesh	74	67
Cambodia	41	24
PRC	77	71
India	86	82
Indonesia	77	71
Kazakhstan	86	82
Korea, Rep. of	92	90
Kyrgyz Republic	77	71
Lao PDR	51	37
Malaysia	100	100
Mongolia	62	51
Myanmar	78	72
Nepal	90	87
Pakistan	91	88
Papua New Guinea	39	22
Philippines	85	81
Sri Lanka	79	73
Tajikistan	59	47
Thailand	99	99
Turkmenistan	72	64
Uzbekistan	82	77
Viet Nam	85	81

Highest Access: 100
Lowest: 22 (Ethiopia)

Annex 3: USE: Domestic Consumption of Water (LPCD) — Final Estimate

Country	Estimate 1 1987–2002 Based on WDI-06	Estimates 2 and 3 (Based on WRI) 1987–95	Estimate for 2000	Average	Use Index
Azerbaijan	311	299		305	242.67
Bangladesh	55	44	47	49	−21.82
Cambodia	21	9	17	16	−55.91
PRC	100	60	95	85	15.52
India	153	81	139	124	55.81
Indonesia	94	67	86	82	12.65
Kazakhstan	123	111	123	119	50.19
Korea, Rep. of	408	378		393	333.23
Kyrgyz Republic	180	182		181	114.59
Lao PDR	70	57		64	−6.63
Malaysia	202	191		196	130.07
Mongolia	101	100		100	31.27
Myanmar	20	20		20	−51.77
Nepal	41	38		40	−31.19
Pakistan	74	69	65	70	−0.38
Papua New Guinea	6	22		14	−57.52
Philippines	193	178	176	182	115.42
Sri Lanka	40	31	37	36	−34.95
Tajikistan	229	230		229	164.19
Thailand	121	82		101	32.42
Turkmenistan	305	16		161	93.45
Uzbekistan	352	288		320	257.53
Viet Nam	218	89	200	169	102.20

Minimum: 70 (Based on the Indian Government Norm)

Maximum: 167 based on the following Singapore data (in LPCD)

1995	1996	1997	1998	1999	2000	2001	2002	Average
172	170	170	166	165	165	165	165	167

LPCD = litres per capita per day, WDI = World Development Indicators, WRI = World Resources Institute.

Annex 3A: Domestic Consumption Estimate 1

	1987–2002 (WDI)				
Country	*Population average 1990–2000*	*billion cubic meters*	*% for domestic use*	*Cubic meters per capita*	*LPCD*
Azerbaijan	7.61	17.3	5	2,273	311
Bangladesh	118.4	79.4	3	671	55
Cambodia	10.59	4.1	2	387	21
PRC	1,205.35	630.3	7	523	100
India	927	645.8	8	697	153
Indonesia	192.61	82.8	8	430	94
Kazakhstan	15.62	35	2	2,241	123
Korea, Rep. of	44.94	18.6	36	414	408
Kyrgyz Republic	4.61	10.1	3	2,189	180
Lao PDR	4.69	3	4	640	70
Malaysia	20.8	9	17	433	202
Mongolia	2.28	0.4	21	175	101
Myanmar	45.45	33.2	1	730	20
Nepal	20.34	10.2	3	501	41
Pakistan	124.74	169.4	2	1,358	74
Papua New Guinea	4.44	0.1	10	23	6
Philippines	68.94	28.5	17	413	193
Sri Lanka	17.37	12.6	2	726	40
Tajikistan	5.74	12	4	2,089	229
Thailand	59.04	87.1	3	1,475	121
Turkmenistan	4.44	24.7	2	5,565	305
Uzbekistan	22.7	58.3	5	2,568	352
Viet Nam	71.83	71.4	8	994	218

LPCD = litres per capita per day, WDI = World Development Indicators.

Annex 3B: Domestic Consumption Estimate 2 (Based on WRI Data)

	Year	Per capita (cubic meters)	% for domestic use	LPCD total	LPCD domestic
Azerbaijan	1995	2,186	5	5,989	299
Bangladesh	1990	134	12	367	44
Cambodia	1987	66	5	181	9
PRC	1993	439	5	1,203	60
India	1990	588	5	1,611	81
Indonesia	1990	407	6	1,115	67
Kazakhstan	1993	2,019	2	5,532	111
Korea, Rep. of	1994	531	26	1,455	378
Kyrgyz Republic	1994	2,219	3	6,079	182
Lao PDR	1987	260	8	712	57
Malaysia	1995	633	11	1,734	191
Mongolia	1993	182	20	499	100
Myanmar	1987	102	7	279	20
Nepal	1994	1,397	1	3,827	38
Pakistan	1991	1,267	2	3,471	69
Papua New Guinea	1987	28	29	77	22
Philippines	1995	811	8	2,222	178
Sri Lanka	1990	573	2	1,570	31
Tajikistan	1994	2,095	4	5,740	230
Thailand	1990	596	5	1,633	82
Turkmenistan	1994	597	1	1,636	16
Uzbekistan	1994	2,626	4	7,195	288
Viet Nam	1990	814	4	2,230	89

LPCD = litres per capita per day, WRI = World Resources Institute.

Annex 3C: Domestic Consumption Estimate 3 (Based on WRI Data)

Country	Withdrawal Per capita cubic meters, 2000	Domestic Use (%)	Total Consumption LPCD	Domestic Consumption LPCD
Bangladesh	576	3	1,578	47
Cambodia	311	2	852	17
PRC	494	7	1,353	95
India	635	8	1,740	139
Indonesia	391	8	1,071	86
Kazakhstan	2,238	2	6,131	123
Korea, Rep. of				
Kyrgyz Republic				
Lao PDR				
Malaysia				
Mongolia				
Myanmar				
Nepal				
Pakistan	1,187	2	3,252	65
Papua New Guinea				
Philippines	377	17	1,033	176
Sri Lanka	678	2	1,857	37
Tajikistan				
Thailand				
Turkmenistan				
Uzbekistan				
Viet Nam	914	8	2,504	200

LPCD = litres per capita per day, WRI = World Resources Institute.

Annex 4: Capacity

Country	GDP/capita PPP $ 2004	Log	Capacity Index
Azerbaijan	3,810	3.58	52
Bangladesh	1,970	3.29	33
Cambodia	2,310	3.36	37
PRC	5,890	3.77	64
India	3,120	3.49	46
Indonesia	3,480	3.54	49
Kazakhstan	6,930	3.84	69
Korea, Rep. of	20,530	4.31	100
Kyrgyz Republic	1,860	3.27	31
Lao PDR	1,880	3.27	31
Malaysia	9,720	3.99	79
Mongolia	2,040	3.31	34
Myanmar	1,550	3.19	26
Nepal	1,480	3.17	25
Pakistan	2,170	3.33	36
Papua New Guinea	2,280	3.36	37
Philippines	4,950	3.69	59
Sri Lanka	4,210	3.62	55
Tajikistan	1,160	3.06	18
Thailand	7,930	3.90	73
Turkmenistan	1,860	3.27	31
Uzbekistan	1,860	3.27	31
Viet Nam	2,700	3.43	42
NORMS			
Malawi	630	2.80	
Korea, Rep. of	20,530	4.31	

GDP = gross domestic product, PPP = purchasing power parity.

Annex 5: Quality

Country	Diarrhoea Death rate	Quality Index
Azerbaijan	11.1	88.92
Bangladesh	47.4	52.59
Cambodia	85.7	14.26
PRC	8.3	91.68
India	43.5	56.52
Indonesia	16.3	83.65
Kazakhstan	2.3	97.71
Korea, Rep. of	0.5	99.46
Kyrgyz Republic	13.0	86.96
Lao PDR	98.5	1.52
Malaysia	1.3	98.71
Mongolia	33.8	66.23
Myanmar	51.0	48.98
Nepal	68.0	32.03
Pakistan	79.0	21.02
Papua New Guinea	42.9	57.06
Philippines	15.6	84.44
Sri Lanka	3.6	96.41
Tajikistan	32.8	67.20
Thailand	7.5	92.53
Turkmenistan	22.9	77.12
Uzbekistan	2.1	97.88
Viet Nam	13.4	86.61

NORMS

Angola		370.0
Ireland, Lithuania, Portugal, Slovakia, Poland, Italy, Austria, Czech. Republic		0.1

Annex 6: Urban House Connections and Overall Access Rates for Water

Country	Years	Number of Years	House Connections (%)	Access 1994	Access 2004	Average Access
Afghanistan	1997–2003	2	10.85	10	63	36.5
Bangladesh	1994–2004	9	26.5	83	82	82.5
Cambodia	2000–2004	2	34		64	64
PRC	1989–2000	5	82.9	99	93	96
India	1993–2001	6	50.1	89	95	92
Indonesia	1991–2004	11	29.08	92	87	89.5
Kazakhstan	1995–1999	3	88.66	97	97	97
Lao PDR	2000–2003	2	44.3		79	79
Malaysia	2003	1	98	100	100	100
Maldives	1996–2001	2	76.55	100	98	99
Mongolia	1996–2000	3	45.5	87	87	87
Myanmar	1995–2003	5	16.88	86	80	83
Nepal	1991–2004	7	48.57	95	96	95.5
Pakistan	1991–2003	9	54.41	95	96	95.5
PNG	1996	1	60.5	88	88	88
Philippines	1993–2003	4	49.5	95	87	91
Solomon Islands	1990	1	76		94	94
Sri Lanka	1987–2004	3	34.33	91	98	94.5
Thailand	1987–2000	3	73.03	98	98	98
Vanuatu	1989–1998	2	77.95	93	86	89.5
Viet Nam	1996–2002	5	55.6	90	99	94.5

Annex 7:   Rural House Connections and Overall Access Rates for Water

Country	Years	Number of Years	House Connections (%)	Access 1990	Access 2004	Average Access
Afghanistan	1997–2003	2	0	3	31	17
Bangladesh	1994–2004	8	0.23	69	72	70.5
Cambodia	2000–2004	2	1.5		35	35
PRC	1989–2000	5	44.1	59	67	63
India	1993–2001	6	8.28	64	83	73.5
Indonesia	1991–2004	11	4.41	63	69	66
Kazakhstan	1995–1999	3	27	73	73	73
Lao PDR	2000–2003	2	5.85		43	43
Malaysia	2003	1	87	96	96	96
Maldives	1996–2001	2	0.1	95	76	85.5
Mongolia	1996–2000	3	0.96	30	30	30
Myanmar	1995–2003	5	1.62	47	77	62
Nepal	1991–2004	7	8.28	67	89	78
Pakistan	1991–2003	9	12.94	78	89	83.5
PNG	1996	1	3.6	32	32	32
Philippines	1993–2003	4	15.5	80	82	81
Sri Lanka	1987–2004	3	3.76	62	74	68
Thailand	1987–2000	3	11.93	94	100	97
Vanuatu	1989–1998	2	27.85	53	52	52.5
Viet Nam	1996–2002	5	2.22	59	80	69.5

Annex 8: IDWA-II

Economy	Resource Index	Use Index	Capacity Index	Quality Index	House Connection %		IDWA II
					Urban	Rural	
Bangladesh	56	−22	33	53	26	0	24
Cambodia	77	−56	37	14	34	2	18
PRC	65	16	64	92	83	44	61
India	60	56	46	57	50	8	46
Indonesia	80	13	49	84	29	4	43
Kazakhstan	72	50	69	98	89	27	67
Lao PDR	88	−7	31	2	44	6	27
Malaysia	85	100	79	99	98	87	91
Mongolia	81	31	34	66	46	1	43
Myanmar	83	−52	26	49	17	2	21
Nepal	75	−31	25	32	48	8	26
Pakistan	49	0	36	21	54	13	29
Papua New Guinea	100	−58	37	57	60	4	33
Philippines	73	100	59	84	49	15	63
Sri Lanka	66	−35	55	96	34	4	37
Thailand	68	32	73	93	73	12	58
Viet Nam	71	102	42	87	56	2	60

3

Global Situation on Drinking Water Adequacy

Ngo Quang Vinh

This paper provides estimates of the Index of Drinking Water Adequacy (IDWA) for 144 countries for the most recent year for which component indicators are available. Inter-correlations between IDWA and selected human development and governance indicators are explored and implications noted.

1. Introduction

The Asian Development Bank released the first ever *Asian Water Development Outlook* (AWDO)[1] in late 2007. The report contained a composite index known as the Index of Drinking Water Adequacy (IDWA), based on an aggregation of five components (see Box 1) and computed for 23 Asian countries. The index was since fine-tuned, christened IDWA II, with the incorporation of water supply via house connections representing access instead of the general access

---

[1] Asian Development Bank (2007).

indicator (see Box 1).[2] Further work on IDWA includes the one pertaining to IDWA for 15 Indian States.[3]

This paper extends IDWA to 144 countries of the world as against the coverage of 23 Asian economies in IDWA-I and IDWA-II. In view of the global coverage, the present variant of IDWA is labeled as IDWA-GLOBAL or IDWA-G for short.

2. Components of Index of Drinking Water Adequacy (IDWA-G)

The 144 countries covered in IDWA-G encompass both indus-trialized and developing countries. The components of IDWA-G are explained below, for the sake of completeness and continuity though there is little difference between the components noted here and those in the earlier exercises (Box 1).

Resources

The index for this component is based on estimates of renewable internal fresh water resources per capita taken from the World Resources Institute (WRI) 2007 and the per capita figures are converted to a log scale. The index R_j for country j is calculated as follows:

$$\text{Resource Indicator for country j} = \left[\frac{(\log R_j - \log R_{min})}{(\log R_{max} - \log R_{min})} \right] \times 100$$

where R_{min} is the internal resource per capita of Egypt and R_{max} is the internal resource per capita of Iceland. The raw data and the resource indicator for each country are in Annex 1.[4]

---

[2] Seetharam, K.E. and Rao, B. (2010). See also paper 4 in this volume.
[3] Seetharam, K.E. and Rao, B. (2009).
[4] Due to the large size of the tables, we have opted to place them in the Annex.

Box 1

Components of IDWA-I and IDWA-II

Resources: Estimates of renewable internal fresh water resources per capita are taken from the World Development Indicators (WDI) 2006, which refer to 2004.

Access: The latest (2004) estimates of access are measured as percent of population with access to a sustainable 'improved' water source. [This access indicator was used in the original version of IDWA since known as IDWA-I, while IDWA-II uses house connections as the access parameter.]

Capacity to Buy Water: Per capita GDP in PPP dollars is used as a measure of a nation's capacity to produce/purchase and supply adequate amounts of drinking water.

Use: This component calculates the per capita water consumption by the domestic sector, which is referred to as 'drinking water' in this paper.

Quality: WHO data on diarrheal deaths per 100,000 people for the year 2000 are used as an indirect measure of drinking water quality.

Access (House connection)

In IDWA-II, urban house connections and rural house connections were calculated and used separately, which implies that each had an equal weight. The refinement in IDWA-G is to take a weighted average of the rural and urban access rates, the weights being the respective population proportions.

As for data sources, the house connections rates by country are from the Water and Sanitation Information Website of the Joint Monitoring Program of the WHO and the UN Children's Fund. Rural and urban population percentages are from the Population Reference Bureau website.[5]

---

[5] See <http://www.prb.org/Datafinder/Topic/Bar.aspx?sort=v&order=d&variable=123>.

House connection indicator for each country is available in Annex 2.

Capacity

Similar to IDWA-I and IDWA-II, per capita GDP in PPP US dollars is used as a measure of a nation's capacity to produce/purchase adequate amount of drinking water. For minimum-maximum estimates from among the 144 countries, the respective figures are US$281 (Dem. Rep. of Congo) and US$50,078 (Norway).

$$\text{Capacity indicator for country } j = \left[\frac{(\log C_j - \log C_{min})}{(\log C_{max} - \log C_{min})} \right] \times 100$$

Annex 3 provides capacity indicator for each country.

Use

The computation of this index is rather challenging as noted in the earlier work on IDWA. Based on three different sets of data from WDI and WRI we obtain the annual use per capita and then the annual figure is converted into litres per capita per day (LPCD). The norms for minimum and maximum consumption levels are retained as 70 and 167 LPCD as in IDWA-I and IDWA-II, and the use indicator for country j is calculated as $\left[\frac{(U_j - 70)}{(167 - 70)} \right] \times 100$.

In IDWA-I, some computed use indicators were reset to 100 if they were higher than 100 and negative values were left untouched. Given the very small number of countries in total and those with negative indicators, the variation was not huge and hence did not create a major bias on that score. In a global context with 144 countries, this method could adversely affect the extent of variability of final IDWA. In the actual data set, it is found that the minimum value is −67 LPCD, maximum value 774 LPCD, median value 61.7 LPCD and the average value 116 LPCD. This is a relatively large extent of variation and could potentially dominate other indexes, which are all computed on a scale 0 and 100.

To partly smooth the series, log values are preferred to the raw numbers. Negative values are reset to 1 to ensure (in log form) zero for the minimum. Log of the maximum value has been set at 100 and other values are prorated between the minimum and the maximum. Details are in Annex 4.

Quality

Previously, IDWA used data on diarrheal death rate (expressed as diarrheal deaths per 100,000 people for the year 2000) as a simple proxy for water quality. One of the drawbacks in using diarrheal death rate is that bad water quality cannot always be fatal. Therefore, it is logical to incorporate as many other water related diseases as possible in our calculation and come up with integrated morbidity cum mortality measure. Thus, one could consider using DALY (see Box 2) which is widely accepted by the World Health Organization (WHO) and the World Bank (WB) and readily available in the WHO's database.

Box 2

Disability-adjusted-life year (DALY)

DALY is a summary measure of the health status of a population, integrating mortality with morbidity and disability information into a single unit. Originally conceptualized in Murray (1994), it was firstly developed as an input into the World Bank's *World Development Report 1993: Investing in Health.* Currently, DALY is widely being used as a tool for policy-making in a wide range of countries and among international organizations' documents.

One DALY represents the loss of one healthy year of life. For each disease DALYs are calculated as the sum of the years of life lost (YLL) due to premature mortality and the years of productive life lost due to disability (YLD) for incident cases of the ill-health condition in question.

DALYs = YLL + YLD where:

YLL = N*L (N = number of deaths; L = standard life expectancy at age of death in years), and

YLD = I*DW*L (I = number of incident cases; DW = disability weight; L = average duration of the case until remission or death in years)

Composite DALY for 10 out of 25 water and sanitation related diseases (Annex 5A) in the WHO list[6] have been collected and then juxtaposed with respective data on diarrheal death rate (Annex 5B). Surprisingly, the correlation coefficient between DALY based on the ten water and sanitation related diseases and the diarrheal death rate is quite significant at 0.941. As a result, we decided to keep the diarrheal death rate as the proxy for water quality as in the previous IDWA papers.

$$\text{Quality indicator based on diarrheal death rate for country j} = \frac{(Q_{max} - Q_j)}{(Q_{max} - Q_{min})} \times 100.$$

3. IDWA-G: Trends and Implications

Taking the average of the aforementioned five indicators for each country, we have IDWA-G for 144 countries (Annex 6). Following the suggestions in the papers on IDWA-I and IDWA-II referred to earlier, we explore the relationships between IDWA-G and key indicators of human development. We also explore the relationships between IDWA-G and indicators of governance (Box 3 and Box 4).

---

[6] See <http://www.who.int/water_sanitation_health/diseases/diseasefact/en/print.html>.

Box 3

Worldwide Governance Indicators (WGI)[7]

WGI covers 212 countries/territories and measures six dimensions of governance between 1996 and 2008: Voice and Accountability, Political Stability and Absence of Violence/Terrorism, Government Effectiveness, Regulatory Quality, Rule of Law, and Control of Corruption. These aggregate indicators are based on hundreds of specific and disaggregated individual variables measuring various dimensions of governance taken from 35 data sources provided by 33 organizations. The data reflect the views on governance of public sector, private sector and NGO experts, as well as thousands of citizen and firm survey respondents worldwide. Though WGI is compiled by the World Bank experts and supported by the World Bank, its findings, interpretation and conclusions do not necessarily represent the views of the World Bank.

In order to explore the relationship between IDWA-G and WGI, we sum up the indicators for six dimensions of WGI and take the average value to compute the correlation coefficient.

Box 4

Corruption Perceptions Index (CPI)[8]

The CPI is developed by Transparency International and designed to score the perceived level of public-sector corruption in a country/territory. The CPI is based on 13 independent surveys and gives a score to each country/territory between 0 and 10. The higher the score, the cleaner the government of a country/territory is. The CPI (2009) used in this paper consists of only 142 countries due to data availability.

---

[7] Kaufmann *et al.* (2009).

[8] See <http://www.transparency.org/policy_research/surveys_indices/cpi/2009/cpi_2009_table>.

Not unexpectedly, HDI and IDWA-G have a high correlation coefficient of 0.93. The evidence in Table 1 clearly brings out the stability of the IDWA-HDI linkage. Similar comment applies to the HPI-IDWA relationship as well. This lends strong support to argue a case for investing relatively more in the water sector.

Table 1: Correlation coefficients between IDWA-G and other indexes

	144 Countries	*144 less OECD Countries*	*Asia without OECD Countries*	*Non-Asia without OECD Countries*
HDI	0.93	0.91 *(118)*	0.84 *(32)*	0.93 *(87)*
WGI	0.70	0.54 *(118)*	0.45 *(32)*	0.57 *(87)*
CPI	0.64	0.45 *(115)*	0.40 *(32)*	0.48 *(85)*
Human Poverty Index (HPI)[9]		−0.85 *(112)*	−0.87 *(31)*	−0.85 *(80)*

Notes: 1. The figures in the parentheses refer to the number of countries in the sample.

2. As the computation method for Human Poverty Indexes (HPI) for OECD members is not identical to that for developing countries, we have opted not to calculate the correlation coefficient between IDWA-G and HPI for all 145 countries. The correlation coefficient between IDWA-G and HPI for selected OECD countries turns out to be low at −0.14.

The correlation coefficients between IDWA-G and WGI and CPI are relatively low but not too low at 0.70 and 0.64, respectively. Good governance and clean government help improve IDWA-G. Yet it can be seen that it is the presence of the OECD nations that lends strength to both the sets of correlations, and hence one must not read too much into them despite common sense dictating the efficacy of water delivery when governance structures are of good standing.

---

[9] See *Technical Note 1, Human Development Report 2007/2008*, United Nations Development Program, for more details (<http://hdr.undp.org/en/media/HDR_20072008_Tech_Note_1.pdf>).

4. Concluding Notes

Between IDWA-II reported in the earlier paper on Asian economies and IDWA-G reported here for 144 economies spanning the entire world, there are bound to be significant differences. These are brought out in the data in Table 2 at the aggregate index level and at component level.

Table 2: IDWA-II and IDWA-G for China and India

	China	*India*
IDWA-II	*61*	*46*
IDWA-G	*63*	*53*
Resource (IDWA-II)	65	60
Resource (IDWA-G)	45	38
Access (IDWA-II) (Urban/Rural)	83/44	50/8
Access (IDWA-G)	69*	31**
Capacity (IDWA-II)	64	46
Capacity (IDWA-G)	54	42
Use (IDWA-II)	16	56
Use (IDWA-G)	49	64
Quality (IDWA-II)	92	57
Quality (IDWA-G)	98	88

Notes: * Based on rural and urban population proportions of 0.46 and 0.54 and respective house connection rates of 85% and 55%.

** Based on rural and urban population proportions of 0.29 and 0.71 and respective house connection rates of 60% and 19%.

For the differences between IDWA-II and IDWA-G, three reasons can be found. First, the databases differ — regional versus global. Second and arising from the first, reference norms [minima and maxima] differ. For instance, for resource [cu m per capita per year], IDWA-G used the maximum observed for Iceland [566,667 cu m] and the minimum observed for Egypt [24 cu m] for the year 2007. In contrast, in IDWA-II, the respective figures used were for Papua New Guinea [138,775 cu m for 2004] and a hypothetical number [1 cu m]. Finally, in IDWA-II, urban house connection rates and rural house connection rates were used as independent components, which imply equal weights for urban and rural.

A key implication arising out of the simple study of the differences between IDWA-II and IDWA-G is that one must not hesitate to improve upon the global IDWA as new and better data become available. Towards this end, the Institute of Water Policy, Lee Kuan Yew School of Public Policy, National University of Singapore plans to annually update IDWA-G and in its stride also look into up-to-date data as well as possible methodological improvements.

References

Anand, S. and K. Hanson (1997). Disability-adjusted Life Years: A Critical Review, *Journal of Health Economics*, 16, 685–702.

Asian Development Bank (2007). *Asian Water and Development Outlook, 2007*. Manila: Asian Development Bank.

Biswas, A.K. and Seetharam, K.E. (2008). Achieving Water Security an Asia, *International Journal of Water Resources Development*.

Kaufmann, D., A. Kraay, and M. Mastruzzi (2009). Governance Matters III: Aggregate and Individual Governance Indicators 1996–2008, *The World Bank Policy Research Working Paper* 4978.

Murray, C.J.L. (1994). Quantifying the Burden of Disease: The Technical Basis for Disability-adjusted Life Years, *Bulletin of the World Health Organization*, 72, 429–55.

Population Reference Bureau (2009). *World Population Data Sheet*. Available at <http://www.prb.org/Datafinder/Topic/Bar.aspx?sort=v&order=d&variable=123> [Accessed February 28, 2010].

Seetharam, K.E. and B. Rao (2009). Index of Drinking Water Adequacy (IDWA) for the States of India, *Journal of Infrastructure Development*, 1, 179–92.

———— (2010). Index of Drinking Water Adequacy for the Asian Economies, *Water Policy*, 12, Supplement 1, 135–54.

The Joint Monitoring Program for Water Supply and Sanitation. *Country profiles: Improved water coverage estimates (1980–2008)*. Available at <http://www.wssinfo.org/resources/documents.html?type=country_files> [Accessed February 28, 2010].

Transparency International (2009). *Corruption Perceptions Index*. Available at <http://www.transparency.org/policy_research/surveys_indices/cpi/2009/cpi_2009_table> [Accessed February 28, 2010].

United Nation's Development Programme (2008). *Technical Note 1, Human Development Report 2007/2008.* Available at <http://hdr.undp.org/en/media/HDR_20072008_Tech_Note_1.pdf> [Accessed February 28, 2010].

World Health Organization. *Water and Sanitation Related Diseases Fact Sheets.* Available at <http://www.who.int/water_sanitation_health/diseases/diseasefact/en/print.html> [Accessed February 28, 2010].

Annexes

Annex 1: Resource Index (Resource is given as cubic metres per capita per year, 2007)

Country	Resource	Log value of resource	Resources index
Albania	8,504.6	9.05	58
Algeria	332.2	5.81	26
Angola	8,774.5	9.08	59
Antigua and Barbuda	619	6.43	32
Argentina	6,981.9	8.85	56
Armenia	3,024.7	8.01	48
Australia	23,911.4	10.08	69
Austria	6,692.6	8.81	56
Azerbaijan	950.7	6.86	37
Bangladesh	714	6.57	34
Belarus	3,856.9	8.26	51
Belgium	1,148	7.05	39
Belize	57,142.9	10.95	77
Benin	1,148.1	7.05	39
Bhutan	42,035.4	10.65	74
Bolivia	31,866.8	10.37	71
Botswana	1,369.1	7.22	40
Brazil	28,315.9	10.25	70
Bulgaria	2,757.4	7.92	47
Burkina Faso	890.2	6.79	36
Burundi	1,235.7	7.12	39
Cambodia	8,236.8	9.02	58
Cameroon	16,178.7	9.69	65
Canada	86,752.7	11.37	81
Cape Verde	566	6.34	32
Cote d'Ivoire	4,093.8	8.32	51
Central African Rep	33,967.7	10.43	72

continued overleaf

Annex 1: Continued

Country	Resource	Log value of resource	Resources index
Chad	1,455.9	7.28	41
Chile	53,141	10.88	77
China	2,112.4	7.66	45
Colombia	44,982.1	10.71	75
Comoros	1,426.9	7.26	41
Congo	52,383.2	10.87	76
Congo, Dem Rep	14,712.1	9.60	64
Costa Rica	25,156.7	10.13	69
Cyprus	913.3	6.82	36
Czech Rep	1,289.5	7.16	40
Denmark	1,098.7	7.00	38
Djibouti	365.9	5.90	27
Dominican Rep	2,295	7.74	45
Ecuador	31,739	10.37	71
Egypt	23.4	3.15	0
El Salvador	2,494.4	7.82	46
Equatorial Guinea	49,335.9	10.81	76
Eritrea	594.7	6.39	32
Estonia	9,623	9.17	60
Ethiopia	1,502.9	7.32	41
Fiji	33,159.1	10.41	72
Finland	20,288.2	9.92	67
France	2,929.1	7.98	48
Gabon	114,765.6	11.65	84
Gambia	1,882.1	7.54	43
Georgia	13,223.4	9.49	63
Germany	1,293.4	7.17	40
Ghana	1,317.7	7.18	40
Greece	5,197.1	8.56	54
Guatemala	8,254	9.02	58
Guinea	23,042.4	10.05	68
Guinea-Bissau	9,512.5	9.16	60
Guyana	320,478.7	12.68	94
Haiti	1,483	7.30	41
Honduras	12,754.8	9.45	62
Hungary	597.3	6.39	32
Iceland	566,666.7	13.25	100
India	1,110	7.01	38
Indonesia	12,440.8	9.43	62

Annex 1:   Continued

Country	Resource	Log value of resource	Resources index
Iran, Islamic Rep	1,804.3	7.50	43
Ireland	11,483.5	9.35	61
Israel	107.7	4.68	15
Italy	3,137.2	8.05	49
Jamaica	3,519.5	8.17	50
Japan	3,350.9	8.12	49
Jordan	114	4.74	16
Kazakhstan	5,095.3	8.54	53
Kenya	574.8	6.35	32
Korea, Rep	1,347.1	7.21	40
Kyrgyzstan	8,624.2	9.06	59
Lao People's Dem Rep	30,747.6	10.33	71
Latvia	7,329.2	8.90	57
Lebanon	1,314	7.18	40
Lesotho	2,930	7.98	48
Liberia	57,937.4	10.97	77
Libyan Arab Jamahiriya	98.6	4.59	14
Lithuana	4,572.4	8.43	52
Madagascar	17,186	9.75	65
Malawi	1,199.8	7.09	39
Malaysia	22,103.7	10.00	68
Maldives	86.7	4.46	13
Mali	4,188.5	8.34	51
Malta	124.7	4.83	17
Mauritania	123.2	4.81	16
Mauritius	2,171.3	7.68	45
Mexico	3,732	8.22	50
Moldova, Rep	238.9	5.48	23
Mongolia	12,836.6	9.46	62
Morocco	894.7	6.80	36
Mozambique	4,887.4	8.49	53
Namibia	2,973	8.00	48
Nepal	7,021.9	8.86	57
Netherlands	669.5	6.51	33
New Zealand	79,892.5	11.29	81
Nicaragua	33,200.3	10.41	72
Niger	234.8	5.46	23
Nigeria	1,610.3	7.38	42
Norway	81,886.4	11.31	81
Pakistan	318.4	5.76	26

continued overleaf

Annex 1: Continued

Country	Resource	Log value of resource	Resources index
Panama	44,098.1	10.69	75
Papua New Guinea	131,010.8	11.78	85
Paraguay	14,584.9	9.59	64
Peru	56,117	10.94	77
Philippines	5,577.3	8.63	54
Romania	1,963.4	7.58	44
Portugal	3,587.3	8.19	50
Russian Federation	30,392.5	10.32	71
Rwanda	1,006.1	6.91	37
Saudi Arabia	93	4.53	14
Senegal	2,111.6	7.66	45
Sierra Leone	27,576.7	10.22	70
South Africa	939.2	6.85	37
Spain	2,550.2	7.84	46
Sri Lanka	2,372.1	7.77	46
Sudan	793.8	6.68	35
Suriname	193,406.6	12.17	89
Swaziland	2,575.6	7.85	47
Sweden	18,801.5	9.84	66
Switzerland	5,553.3	8.62	54
Syrian Arab Rep	350.2	5.86	27
Tajikistan	9,922.2	9.20	60
Tanzania	2,114.9	7.66	45
Thailand	3,216.8	8.08	49
Togo	1,777.4	7.48	43
Trinidad and Tobago	2,924.6	7.98	48
Tunisia	406.5	6.01	28
Turkey	3,020.2	8.01	48
Uganda	1,260.3	7.14	39
Ukraine	1,166.8	7.06	39
United Kingdom	2,415.9	7.79	46
United States	9,215	9.13	59
Uruguay	16,813.9	9.73	65
Uzbekistan	597	6.39	32
Venezuela	26,096.3	10.17	70
Viet Nam	4,239.7	8.35	52
Yemen	183.7	5.21	20
Zambia	6,652.3	8.80	56

Notes: Highest value: 566666.7 (Iceland)
　　　　 Lowest value: 23.4 (Egypt)
Source: World Resources Institute (<http://www.wri.org>).

Annex 2: Access (House Connection-HC) Index

Country	Urban Pop. (%)	Rural Pop. (%)	Urban HC(%)	Rural HC(%)	Nationwide HC(%)
Albania	49	51	95.1	47.6	71
Algeria	63	37	81	21	59
Angola	57	43	22.4	0	13
Antigua and Barbuda	31	69	90	79	82
Argentina	91	9	72	27	68
Armenia	64	36	96.6	71.9	88
Australia	83	17	100	100	100
Austria	67	33	100	100	100
Azerbaijan	52	48	73.6	18.2	47
Bangladesh	25	75	29	0	7
Belarus	74	26	88.8	25.2	72
Belgium	97	3	100	90	100
Belize	51	49	100	63	82
Benin	41	59	49	0	20
Bhutan	31	69	53.7	9.2	23
Bolivia	65	35	87	30	67
Botswana	60	40	82	9	53
Brazil	84	16	91	20	80
Bulgaria	71	29	96.5	73.8	90
Burkina Faso	16	84	15	0	2
Burundi	10	90	51	3	8
Cambodia	15	85	30	1	5
Cameroon	57	43	26.4	0.8	15
Canada	81	19	100	38	88
Cape Verde	59	41	57	37	49
Cote d'Ivoire	48	52	10	2	6
Central African Republic	38	62	8	0	3
Chad	27	73	10	0	3
Chile	87	13	95	45	89
China	46	54	85	55	69
Colombia	75	25	89	46.4	78
Comoros	28	72	48	21	29
Congo	60	40	39.5	2.6	25
Congo, Dem. Rep. of	33	67	53	0	17
Costa Rica	59	41	98.9	91.3	96

continued overleaf

Annex 2:   Continued

Country	Urban Pop. (%)	Rural Pop. (%)	Urban HC (%)	Rural HC (%)	Nationwide HC (%)
Cyprus	62	38	100	100	100
Czech Republic	74	26	97	91	95
Denmark	72	28	100	100	100
Djibouti	87	13	31	1	27
Dominican Republic	64	36	83	48	70
Ecuador	63	37	77	42	64
Egypt	43	57	98.6	73.6	84
El Salvador	60	40	83	37.8	65
Equatorial Guinea	39	61	15.5	0	6
Eritrea	21	79	42	0.1	9
Estonia	69	31	93	68	85
Ethiopia	16	84	73	4	15
Fiji	51	49	32	7	20
Finland	63	37	99	91	96
France	77	23	100	95	99
Gabon	84	16	52.2	8.4	45
Gambia	54	46	75	0	41
Georgia	53	47	91.2	30	62
Germany	73	27	100	97	99
Ghana	48	52	34.6	5.7	20
Greece	60	40	99.8	96.2	98
Guatemala	47	53	79	34	55
Guinea	33	67	29	0	10
Guinea-Bissau	30	70	20	0	6
Guyana	28	72	80	47.5	57
Haiti	43	57	15	24	20
Honduras	49	51	89	60.5	74
Hungary	66	34	96	88	93
Iceland	93	7	100	100	100
India	29	71	60	19	31
Indonesia	43	57	31	4.6	16
Iran	67	33	86	74	82
Ireland	60	40	98.7	96.2	98
Israel	92	8	100	98	100
Italy	68	32	100	96	99
Jamaica	52	48	92	46	70
Japan	86	14	97	97	97

Annex 2: Continued

Country	Urban Pop. (%)	Rural Pop. (%)	Urban HC (%)	Rural HC (%)	Nationwide HC (%)
Jordan	83	17	98	70	93
Kazakhstan	53	47	87	23	57
Kenya	19	81	58	12	21
Korea, South	82	18	96	39	86
Kyrgyzstan	35	65	85	33	51
Laos	27	73	53	7	19
Latvia	68	32	92.4	58.7	82
Lebanon	87	13	100	85	98
Lesotho	24	76	37	2	10
Liberia	58	42	4	1	3
Libya	77	23	54.2	54.9	54
Lithuana	67	33	92.1	55.6	80
Madagascar	30	70	18	0	5
Malawi	17	83	61	0	10
Malaysia	68	32	98	87	94
Maldives	35	65	100	0	35
Mali	31	69	74	70	71
Malta	94	6	100	96	100
Mauritania	40	60	86	41	59
Mauritius	42	58	80	82	81
Mexico	77	23	95	72	90
Moldova	41	59	78	9	37
Mongolia	60	40	57	0	34
Morocco	56	44	84	6	50
Mozambique	29	71	18	2	7
Namibia	35	65	82	0	29
Nepal	17	83	30	0	5
Netherlands	66	34	100	100	100
New Zealand	86	14	100	82	97
Nicaragua	58	42	88	14	57
Niger	17	83	35	1	7
Nigeria	47	53	16	15	15
Norway	80	20	100	100	100
Pakistan	35	65	42	24	30
Panama	64	36	94.1	77.3	88
Papua New Guinea	13	87	60.5	3.6	11
Paraguay	57	43	69	11	44

continued overleaf

Annex 2:   Continued

Country	Urban Pop. (%)	Rural Pop. (%)	Urban HC (%)	Rural HC (%)	Nationwide HC (%)
Peru	76	24	76	29	65
Philippines	63	37	47	15	35
Portugal	55	45	98.4	97	98
Romania	55	45	86	8	51
Russia	73	27	81	54	74
Rwanda	18	82	59	3	13
Saudi Arabia	81	19	89	26	77
Senegal	41	59	74	23	44
Sierra Leone	37	63	29.7	1.2	12
South Africa	59	41	87	47	71
Spain	77	23	99.2	99.6	99
Sri Lanka	15	85	67	13	21
Sudan	38	62	69	19	38
Suriname	67	33	90.6	48.4	77
Swaziland	24	76	47.4	16.6	24
Sweden	84	16	100	100	100
Switzerland	73	27	100	99	100
Syria	54	46	92	65	80
Tajikistan	26	74	82	26	41
Tanzania	25	75	18	1	5
Thailand	36	64	76	2	29
Togo	40	60	19	1	8
Trinidad and Tobago	12	88	81	67	69
Tunisia	66	34	96	37	76
Turkey	63	37	93	80	88
Uganda	13	87	10	2	3
Ukraine	68	32	85	47	73
United Kingdom	80	20	99.5	98	99
United States	79	21	97	46	86
Uruguay	94	6	97	84	96
Uzbekistan	36	64	83	29	48
Venezuela	88	12	88	51	84
Vietnam	28	72	85	4	27
Yemen	29	71	61	12	26
Zambia	37	63	47	2	19

Source: Population Reference Bureau (<www.prb.org>) and Joint Monitoring Program
    WHO/UNICEF (<www.wssinfo.org>).

Annex 3: Capacity (GDP per capita, PPP, current international US$ 2006)

Country	GDP per capita PPP	Log value	Capacity index
Albania	5,886	8.68	58.7
Algeria	6,347	8.76	60.1
Angola	4,434	8.40	53.2
Antigua and Barbuda	16,578	9.72	78.7
Argentina	11,985	9.39	72.4
Armenia	4,879	8.49	55.1
Australia	35,547	10.48	93.4
Austria	36,049	10.49	93.7
Azerbaijan	6,280	8.75	59.9
Bangladesh	1,155	7.05	27.3
Belarus	9,732	9.18	68.4
Belgium	33,543	10.42	92.3
Belize	7,846	8.97	64.2
Benin	1,263	7.14	29.0
Bhutan	4,010	8.30	51.3
Bolivia	3,937	8.28	50.9
Botswana	12,508	9.43	73.2
Brazil	8,949	9.10	66.8
Bulgaria	10,274	9.24	69.4
Burkina Faso	1,130	7.03	26.8
Burundi	333	5.81	3.3
Cambodia	1,619	7.39	33.8
Cameroon	2,089	7.64	38.7
Canada	36,713	10.51	94.0
Cape Verde	2,697	7.90	43.6
Cote d'Ivoire	1,650	7.41	34.2
Central African Rep	690	6.54	17.3
Chad	1,478	7.30	32.0
Chile	13,030	9.48	74.0
China	4,644	8.44	54.1
Colombia	6,378	8.76	60.2
Comoros	1,144	7.04	27.1
Congo	3,487	8.16	48.6
Congo, Dem Rep	281	5.64	0.0
Costa Rica	9,564	9.17	68.1
Cyprus	25,882	10.16	87.3

continued overleaf

Annex 3: Continued

Country	GDP per capita PPP	Log value	Capacity index
Czech Rep	22,118	10.00	84.2
Denmark	35,692	10.48	93.5
Djibouti	1,966	7.58	37.5
Dominican Rep	5,866	8.68	58.6
Ecuador	7,145	8.87	62.4
Egypt	4,953	8.51	55.4
El Salvador	5,765	8.66	58.3
Equatorial Guinea	27,161	10.21	88.2
Eritrea	682	6.53	17.1
Estonia	18,969	9.85	81.3
Ethiopia	636	6.46	15.8
Fiji	4,548	8.42	53.7
Finland	33,022	10.40	92.0
France	31,992	10.37	91.4
Gabon	14,208	9.56	75.7
Gambia	1,130	7.03	26.8
Georgia	4,010	8.30	51.3
Germany	32,322	10.38	91.6
Ghana	1,245	7.13	28.7
Greece	31,382	10.35	91.0
Guatemala	5,175	8.55	56.2
Guinea	1,149	7.05	27.2
Guinea-Bissau	478	6.17	10.3
Guyana	3,547	8.17	48.9
Haiti	1,224	7.11	28.4
Honduras	3,543	8.17	48.9
Hungary	18,277	9.81	80.6
Iceland	36,923	10.52	94.1
India	2,469	7.81	41.9
Indonesia	3,454	8.15	48.4
Iran, Islamic Rep	9,906	9.20	68.7
Ireland	40,268	10.60	95.8
Israel	24,097	10.09	85.9
Italy	29,053	10.28	89.5
Jamaica	7,567	8.93	63.5
Japan	31,947	10.37	91.3
Jordan	4,628	8.44	54.1

Annex 3: Continued

Country	GDP per capita PPP	Log value	Capacity index
Kazakhstan	9,832	9.19	68.6
Kenya	1,467	7.29	31.9
Korea, Rep	22,988	10.04	85.0
Kyrgyzstan	1,813	7.50	36.0
Lao People's Dem Rep	1,980	7.59	37.7
Latvia	15,350	9.64	77.2
Lebanon	9,741	9.18	68.4
Lesotho	1,440	7.27	31.5
Liberia	334	5.81	3.3
Libyan Arab Jamahiriya	11,622	9.36	71.8
Lithuana	15,738	9.66	77.7
Madagascar	878	6.78	22.0
Malawi	700	6.55	17.6
Malaysia	12,536	9.44	73.3
Maldives	5,008	8.52	55.6
Mali	1,058	6.96	25.6
Malta	21,720	9.99	83.9
Mauritania	1,890	7.54	36.8
Mauritius	10,571	9.27	70.0
Mexico	12,177	9.41	72.7
Moldova, Rep	2,377	7.77	41.2
Mongolia	2,887	7.97	44.9
Morocco	3,915	8.27	50.8
Mozambique	739	6.61	18.7
Namibia	4,819	8.48	54.8
Nepal	999	6.91	24.5
Netherlands	36,560	10.51	93.9
New Zealand	25,517	10.15	87.0
Nicaragua	2,789	7.93	44.3
Niger	629	6.44	15.5
Nigeria	1,611	7.38	33.7
Norway	50,078	10.82	100.0
Pakistan	2,361	7.77	41.1
Panama	9,255	9.13	67.4
Papua New Guinea	1,817	7.50	36.0
Paraguay	4,034	8.30	51.4
Peru	7,092	8.87	62.3

continued overleaf

Annex 3: Continued

Country	GDP per capita PPP	Log value	Capacity index
Philippines	3,153	8.06	46.6
Portugal	20,784	9.94	83.0
Romania	10,431	9.25	69.7
Russian Federation	13,116	9.48	74.2
Rwanda	738	6.60	18.6
Saudi Arabia	22,296	10.01	84.4
Senegal	1,585	7.37	33.4
Sierra Leone	630	6.45	15.6
South Africa	9,087	9.11	67.1
Spain	28,649	10.26	89.2
Sri Lanka	3,747	8.23	50.0
Sudan	1,931	7.57	37.2
Suriname	7,894	8.97	64.4
Swaziland	4,671	8.45	54.2
Sweden	34,193	10.44	92.6
Switzerland	37,194	10.52	94.3
Syrian Arab Rep	4,225	8.35	52.3
Tajikistan	1,610	7.38	33.7
Tanzania	995	6.90	24.4
Thailand	7,599	8.94	63.6
Togo	776	6.65	19.6
Trinidad and Tobago	17,717	9.78	80.0
Tunisia	6,859	8.83	61.6
Turkey	8,417	9.04	65.6
Uganda	893	6.79	22.3
Ukraine	6,212	8.73	59.7
United Kingdom	33,087	10.41	92.0
United States	43,968	10.69	97.5
Uruguay	10,203	9.23	69.3
Uzbekistan	2,192	7.69	39.6
Venezuela	11,060	9.31	70.9
Viet Nam	2,363	7.77	41.1
Yemen	2,264	7.72	40.3
Zambia	1,259	7.14	28.9

Notes: Highest value: US$ 50,078 (Norway)
        Lowest value: US$ 281 (Congo, Dem. Rep.)
Source: World Resources Institute (<http://www.wri.org>).

Annex 4: Use (Domestic consumption of water, 2000)

Country	Indicator with norms of 70-167 LPCD	Log value	Use index
Albania	353	5.9	88.2
Algeria	51	3.9	59.3
Angola	−56	0.0	0.0
Antigua and Barbuda	62	4.1	62.0
Argentina	308	5.7	86.1
Armenia	739	6.6	99.3
Australia	460	6.1	92.2
Austria	186	5.2	78.6
Azerbaijan	227	5.4	81.6
Bangladesh	−20	0.0	0.0
Belarus	108	4.7	70.5
Belgium	212	5.4	80.6
Belize	50	3.9	58.9
Benin	−56	0.0	0.0
Bhutan	−41	0.0	0.0
Bolivia	−9	0.0	0.0
Botswana	56	4.0	60.6
Brazil	120	4.8	72.0
Bulgaria	39	3.7	55.1
Burkina Faso	−46	0.0	0.0
Burundi	−51	0.0	0.0
Cambodia	−63	0.0	0.0
Cameroon	−38	0.0	0.0
Canada	774	6.7	100.0
Cape Verde	−62	0.0	0.0
Cote d'Ivoire	−34	0.0	0.0
Central African Rep	−56	0.0	0.0
Chad	−59	0.0	0.0
Chile	181	5.2	78.1
China	26	3.2	48.8
Colombia	287	5.7	85.1
Comoros	−53	0.0	0.0
Congo	−46	0.0	0.0
Congo, Dem Rep	−62	0.0	0.0
Costa Rica	486	6.2	93.0
Cyprus	178	5.2	77.9

continued overleaf

Annex 4: Continued

Country	Indicator with norms of 70-167 LPCD	Log value	Use index
Czech Rep	219	5.4	81.0
Denmark	143	5.0	74.6
Djibouti	452	6.1	91.9
Dominican Rep	298	5.7	85.7
Ecuador	396	6.0	89.9
Egypt	157	5.1	76.0
El Salvador	72	4.3	64.3
Equatorial Guinea	493	6.2	93.2
Eritrea	−54	0.0	0.0
Estonia	115	4.7	71.3
Ethiopia	−58	0.0	0.0
Fiji	−38	0.0	0.0
Finland	117	4.8	71.6
France	232	5.4	81.9
Gabon	61	4.1	61.7
Gambia	−57	0.0	0.0
Georgia	360	5.9	88.5
Germany	121	4.8	72.1
Ghana	−39	0.0	0.0
Greece	248	5.5	82.8
Guatemala	−42	0.0	0.0
Guinea	−32	0.0	0.0
Guinea-Bissau	−25	0.0	0.0
Guyana	52	4.0	59.5
Haiti	−55	0.0	0.0
Honduras	−42	0.0	0.0
Hungary	118	4.8	71.7
Iceland	426	6.1	91.0
India	71	4.3	64.1
Indonesia	17	2.9	42.9
Iran, Islamic Rep	145	5.0	74.8
Ireland	121	4.8	72.1
Israel	223	5.4	81.3
Italy	319	5.8	86.7
Jamaica	81	4.4	66.0
Japan	321	5.8	86.8

Annex 4: Continued

Country	Indicator with norms of 70-167 LPCD	Log value	Use index
Jordan	48	3.9	58.3
Kazakhstan	59	4.1	61.4
Kenya	−29	0.0	0.0
Korea, Rep	332	5.8	87.3
Kyrgyzstan	100	4.6	69.3
Lao People's Dem Rep	−8	0.0	0.0
Latvia	116	4.8	71.5
Lebanon	306	5.7	86.1
Lesotho	−41	0.0	0.0
Liberia	−45	0.0	0.0
Libyan Arab Jamahiriya	246	5.5	82.8
Lithuana	168	5.1	77.0
Madagascar	6	1.8	27.3
Malawi	−35	0.0	0.0
Malaysia	116	4.8	71.5
Maldives	−22	0.0	0.0
Mali	71	4.3	64.0
Malta	216	5.4	80.8
Mauritania	91	4.5	67.9
Mauritius	421	6.0	90.8
Mexico	303	5.7	85.9
Moldova, Rep	80	4.4	65.9
Mongolia	27	3.3	49.7
Morocco	50	3.9	58.7
Mozambique	−61	0.0	0.0
Namibia	35	3.6	53.4
Nepal	−37	0.0	0.0
Netherlands	12	2.5	37.9
New Zealand	678	6.5	98.0
Nicaragua	39	3.7	55.0
Niger	−51	0.0	0.0
Nigeria	−32	0.0	0.0
Norway	244	5.5	82.6
Pakistan	−5	0.0	0.0
Panama	454	6.1	92.0
Papua New Guinea	−52	0.0	0.0

continued overleaf

Annex 4: Continued

Country	Indicator with norms of 70-167 LPCD	Log value	Use index
Paraguay	−21	0.0	0.0
Peru	103	4.6	69.7
Philippines	108	4.7	70.4
Portugal	239	5.5	82.3
Romania	194	5.3	79.2
Russian Federation	209	5.3	80.3
Rwanda	−35	0.0	0.0
Saudi Arabia	155	5.0	75.9
Senegal	−48	0.0	0.0
Sierra Leone	−60	0.0	0.0
South Africa	168	5.1	77.0
Spain	249	5.5	83.0
Sri Lanka	−67	0.0	0.0
Sudan	24	3.2	47.7
Suriname	102	4.6	69.6
Swaziland	−15	0.0	0.0
Sweden	276	5.6	84.5
Switzerland	171	5.1	77.3
Syrian Arab Rep	28	3.3	50.3
Tajikistan	147	5.0	75.1
Tanzania	−30	0.0	0.0
Thailand	8	2.1	31.0
Togo	−24	0.0	0.0
Trinidad and Tobago	391	6.0	89.7
Tunisia	37	3.6	54.3
Turkey	161	5.1	76.4
Uganda	−58	0.0	0.0
Ukraine	187	5.2	78.6
United Kingdom	29	3.4	50.7
United States	547	6.3	94.8
Uruguay	8	2.0	30.8
Uzbekistan	261	5.6	83.7
Venezuela	373	5.9	89.0
Viet Nam	133	4.9	73.5
Yemen	−30	0.0	0.0
Zambia	6	1.8	27.2

Note: Maximum value: 774 (Canada)

Source: World Resources Institute (<http://www.wri.org>).

Annex 5A: Raw Data for Estimated DALYs per 100,000 Populations by 10 Water and Sanitation Related Diseases, WHO 2004

Notes:

(1) Diarrheal diseases	(6) Hepatitis C (g)
(2) Ascariasis	(7) Malaria
(3) Japanese encephalitis	(8) Trachoma
(4) Dengue	(9) Onchocerciasis
(5) Hepatitis B (g)	(10) Hookworm disease

Country	(1)	(2)	(3)	(4)	(5)	(6)	(7)	(8)	(9)	(10)	Total
Albania	37	–	–	–	6	3	–	–	–	–	45
Algeria	945	186	–	0	24	11	0	–	–	38	1,204
Angola	12,197	241	–	1	182	82	5,165	31	13	37	17,949
Antigua and Barbuda	132	3	–	–	3	2	–	–	–	4	143
Argentina	130	3	–	–	3	3	1	–	–	3	144
Armenia	114	0	–	–	3	1	17	–	–	0	135
Australia	27	–	–	0	1	1	0	1	–	–	29
Austria	23	–	–	–	2	5	0	–	–	–	30
Azerbaijan	440	0	–	–	14	6	58	–	–	0	520
Bangladesh	1,598	23	17	49	40	14	85	11	–	9	1,847
Belarus	31	–	–	–	2	1	–	–	–	–	34
Belgium	27	–	–	–	3	6	0	–	–	–	36
Belize	506	5	–	0	4	0	40	–	–	3	558
Benin	3,676	238	–	0	127	57	6,857	194	3	36	11,188
Bhutan	2,008	30	34	95	75	30	50	0	–	9	2,331
Bolivia	1,646	55	–	2	51	1	33	–	–	5	1,793
Botswana	751	33	–	0	6	3	323	165	–	39	1,321
Brazil	417	4	–	8	14	10	24	57	0	4	538
Bulgaria	33	–	–	–	7	3	–	–	–	–	44
Burkina Faso	5,682	251	–	0	275	–	7,341	170	1	39	13,758
Burundi	4,632	40	–	0	109	49	2,784	251	60	39	7,965
Cambodia	2,801	156	32	44	182	46	884	18	–	45	4,207
Cameroon	3,053	227	–	0	90	41	4,605	173	148	39	8,376
Canada	28	–	–	–	3	4	0	–	–	–	35
Cape Verde	831	220	–	0	4	2	8	160	–	34	1,258
Cote d'Ivoire	3,295	35	–	1	73	33	5,589	121	19	39	9,204
Central African Rep.	3,959	36	–	0	39	18	5,665	173	223	40	10,152
Chad	4,493	241	–	0	81	37	7,147	252	211	40	12,501
Chile	122	3	–	–	5	3	–	–	–	4	137
China	388	13	22	0	26	11	6	29	–	20	515
Colombia	307	6	–	3	20	3	51	1	0	4	395

continued overleaf

Annex 5A: Continued

Country	(1)	(2)	(3)	(4)	(5)	(6)	(7)	(8)	(9)	(10)	Total
Comoros	1,776	222	–	1	42	19	2,579	199	–	41	4,879
Congo	1,433	37	–	0	25	11	5,091	–	13	37	6,647
Congo, Dem. Rep. of	7,167	40	–	2	31	14	7,091	173	–	43	14,560
Costa Rica	141	4	–	0	5	–	34	–	–	4	189
Cyprus	78	0	–	–	–	–	–	–	–	0	79
Czech Republic	23	–	–	–	1	1	–	–	–	–	25
Denmark	30	–	–	–	2	2	–	–	–	–	33
Djibouti	4,024	223	–	1	87	42	106	165	–	33	4,681
Dominican Republic	571	4	–	89	33	9	20	–	–	3	730
Ecuador	605	46	–	0	4	2	54	–	1	5	717
Egypt	658	7	–	0	50	26	221	175	–	3	1,140
El Salvador	591	4	–	31	–	–	6	–	–	3	636
Equatorial Guinea	3,090	227	–	1	62	28	6,351	150	105	37	10,051
Eritrea	2,181	37	–	1	17	8	2,418	213	–	40	4,913
Estonia	26	–	–	–	3	1	–	–	–	–	30
Ethiopia	3,045	53	–	1	147	66	1,814	267	52	49	5,493
Fiji	201	25	9	11	4	–	–	–	–	42	292
Finland	26	–	–	–	2	3	–	–	–	–	32
France	30	–	–	–	1	3	1	–	–	–	35
Gabon	1,127	220	–	0	15	7	3,300	–	68	43	4,780
Gambia	2,241	213	–	0	86	39	3,378	42	–	38	6,036
Georgia	38	–	–	–	0	0	120	–	–	–	158
Germany	25	–	–	–	7	10	0	–	–	–	41
Ghana	1,609	215	–	0	21	10	4,207	171	1	39	6,272
Greece	23	–	–	–	2	4	–	–	–	–	28
Guatemala	1,021	71	–	13	1	1	68	16	6	4	1,202
Guinea	3,734	227	–	1	66	29	7,206	139	44	43	11,489
Guinea–Bissau	3,652	239	–	1	329	148	6,113	244	2	39	10,767
Guyana	1,155	4	–	–	29	13	193	–	–	4	1,397
Haiti	2,257	54	–	405	247	–	113	–	–	5	3,079
Honduras	908	18	–	57	10	5	269	0	–	3	1,271
Hungary	27	–	–	–	1	2	–	–	–	–	30
Iceland	28	–	–	–	2	1	–	–	–	–	31
India	1,453	24	22	18	41	16	80	13	–	10	1,677
Indonesia	582	22	11	9	26	11	163	0	–	43	867
Iran	473	0	–	–	14	6	99	67	–	0	660
Ireland	27	–	–	–	1	3	0	–	–	–	31
Israel	36	–	–	–	7	12	–	–	–	–	54

Annex 5A: Continued

Country	(1)	(2)	(3)	(4)	(5)	(6)	(7)	(8)	(9)	(10)	Total
Italy	23	–	–	–	8	19	0	–	–	–	50
Jamaica	268	4	–	–	6	3	–	–	–	3	284
Japan	29	–	0	–	12	30	0	–	–	–	71
Jordan	497	0	–	–	15	38	–	–	–	0	550
Kazakhstan	101	–	–	–	17	8	1	–	–	–	127
Kenya	2,592	34	–	1	14	6	2,206	119	–	42	5,015
Korea, South	132	10	13	0	13	6	1	–	–	21	197
Kyrgyzstan	512	1	–	–	41	2	2	–	–	0	559
Laos	3,181	164	84	112	106	43	832	17	–	45	4,585
Latvia	29	–	–	–	4	2	–	–	–	–	34
Lebanon	311	0	–	0	30	15	–	–	–	0	356
Lesotho	2,735	33	–	1	10	5	13	–	–	41	2,839
Liberia	4,910	238	–	0	348	157	7,306	–	196	35	13,190
Libya	334	0	–	0	28	13	0	87	–	0	462
Lithuana	26	–	–	–	3	2	0	–	–	–	32
Madagascar	3,674	232	–	0	36	16	4,856	–	–	43	8,857
Malawi	5,321	42	–	3	147	66	6,184	288	58	45	12,154
Malaysia	117	24	9	17	48	21	28	0	–	41	305
Maldives	615	28	19	1	57	14	125	0	–	9	869
Mali	5,945	255	–	0	222	100	6,545	310	2	54	13,433
Malta	25	–	–	–	5	3	–	–	–	–	32
Mauritania	4,263	222	–	1	52	23	3,569	172	–	37	8,339
Mauritius	119	140	–	1	1	–	1	451	–	35	748
Mexico	249	5	–	0	12	7	5	59	0	3	340
Moldova	52	–	–	–	12	6	–	–	–	–	70
Mongolia	1,194	17	20	0	67	30	–	–	–	19	1,347
Morocco	771	6	–	0	11	5	3	55	–	2	853
Mozambique	5,286	40	–	3	156	70	6,586	214	–	43	12,398
Namibia	1,489	37	–	0	12	5	1,687	–	–	35	3,266
Nepal	2,241	26	22	27	79	21	59	12	–	9	2,497
Netherlands	26	–	–	0	3	2	0	–	–	–	31
New Zealand	30	–	–	–	3	7	–	–	–	–	40
Nicaragua	960	61	–	50	5	2	102	–	–	4	1,184
Niger	7,211	252	–	0	190	85	7,595	222	1	36	15,591
Nigeria	3,624	234	–	1	72	33	6,505	193	271	32	10,964
Norway	32	–	–	–	3	2	1	–	–	–	38
Pakistan	2,604	32	55	13	33	13	113	10	–	9	2,883
Panama	327	7	–	–	32	1	7	–	–	3	377

continued overleaf

Annex 5A: Continued

Country	(1)	(2)	(3)	(4)	(5)	(6)	(7)	(8)	(9)	(10)	Total
Papua New Guinea	1,437	46	41	0	97	20	584	0	–	38	2,265
Paraguay	602	5	–	–	3	1	41	–	–	3	654
Peru	638	47	–	0	46	2	30	–	–	5	768
Philippines	523	27	10	139	14	6	43	–	–	40	802
Portugal	26	–	–	–	6	16	2	–	–	–	51
Romania	47	–	–	–	5	1	–	–	–	–	53
Russia	38	–	–	–	10	4	1	–	–	–	53
Rwanda	5,265	39	–	0	128	57	1,458	172	–	41	7,158
Saudi Arabia	317	0	–	39	19	6	36	108	–	0	525
Senegal	2,516	230	–	1	41	19	4,895	156	6	36	7,899
Sierra Leone	8,772	231	–	3	586	263	7,146	262	76	60	17,398
South Africa	1,069	29	–	0	18	–	8	219	–	37	1,380
Spain	26	–	–	–	5	16	0	–	–	–	47
Sri Lanka	172	19	7	6	10	4	215	–	–	44	477
Sudan	1,992	213	–	1	31	6	2,421	295	30	33	5,023
Suriname	412	4	–	212	12	5	92	–	–	4	741
Swaziland	1,937	49	–	0	41	19	86	140	–	36	2,308
Sweden	25	–	–	–	1	5	–	–	–	–	31
Switzerland	24	–	–	–	4	3	0	–	–	–	31
Syria	451	0	–	0	14	6	1	–	–	0	472
Tajikistan	1,103	0	–	–	60	60	208	–	–	0	1,431
Tanzania	2,919	38	–	1	99	45	5,687	306	–	41	9,136
Thailand	297	19	8	111	135	68	259	0	–	58	954
Togo	2,015	230	–	0	28	13	4,971	154	55	35	7,501
Trinidad and Tobago	111	3	–	–	8	4	–	–	–	4	129
Tunisia	283	0	–	0	20	10	0	–	–	0	313
Turkey	335	–	–	–	39	8	162	–	–	0	545
Uganda	3,954	41	–	1	56	25	6,118	132	86	40	10,454
Ukraine	33	–	–	–	4	2	0	–	–	–	39
United Kingdom	30	–	–	–	2	4	0	–	–	–	36
United States	29	0	–	0	6	26	0	–	–	–	61
Uruguay	132	3	–	–	1	1	–	–	–	3	141
Uzbekistan	147	0	–	–	49	22	0	–	–	0	218
Venezuela	283	10	–	2	9	2	17	–	2	3	328
Vietnam	493	112	14	7	59	28	207	20	–	51	989
Yemen	3,226	9	–	0	18	9	518	37	2	2	3,822
Zambia	4,686	40	–	0	133	60	6,504	157	–	47	11,626

Source: WHO websites <http://www.who.int/water_sanitation_health/diseases/diseasefact/en/print.html>, <http://www.who.int/research/en/>

Annex 5B: Quality indicator

Country	Diarrheal death rate	Quality indicator
Albania	0.17	100.0
Algeria	26.05	93.0
Angola	369.90	0.0
Antigua and Barbuda	0.60	99.8
Argentina	1.13	99.7
Armenia	1.90	99.5
Australia	0.17	100.0
Austria	0.06	100.0
Azerbaijan	11.08	97.0
Bangladesh	47.41	87.2
Belarus	0.24	99.9
Belgium	0.88	99.8
Belize	11.41	96.9
Benin	111.95	69.7
Bhutan	60.39	83.7
Bolivia	48.21	87.0
Botswana	20.51	94.5
Brazil	9.83	97.3
Bulgaria	0.33	99.9
Burkina Faso	172.64	53.3
Burundi	140.76	61.9
Cambodia	85.74	76.8
Cameroon	91.41	75.3
Canada	0.83	99.8
Cape Verde	22.46	93.9
Cote d'Ivoire	101.84	72.5
Central African Republic	122.75	66.8
Chad	136.37	63.1
Chile	1.62	99.6
China	8.32	97.7
Colombia	6.10	98.4
Comoros	51.98	85.9
Congo	43.20	88.3
Congo, Dem. Rep. of	219.15	40.8
Costa Rica	2.92	99.2
Cyprus	0.53	99.9

continued overleaf

Annex 5B: Continued

Country	Diarrheal death rate	Quality indicator
Czech Republic	0.06	100.0
Denmark	1.39	99.6
Djibouti	120.66	67.4
Dominican Republic	14.50	96.1
Ecuador	15.39	95.8
Egypt	18.63	95.0
El Salvador	14.92	96.0
Equatorial Guinea	94.88	74.4
Eritrea	64.75	82.5
Estonia	0.04	100.0
Ethiopia	91.58	75.2
Fiji	6.65	98.2
Finland	0.59	99.8
France	0.93	99.7
Gabon	31.70	91.4
Gambia	67.72	81.7
Georgia	0.68	99.8
Germany	0.33	99.9
Ghana	47.32	87.2
Greece	0.00	100.0
Guatemala	28.14	92.4
Guinea	113.42	69.3
Guinea-Bissau	112.24	69.7
Guyana	40.12	89.2
Haiti	67.63	81.7
Honduras	24.69	93.3
Hungary	0.19	99.9
Iceland	0.39	99.9
India	43.48	88.2
Indonesia	16.35	95.6
Iran	12.75	96.6
Ireland	0.14	100.0
Israel	0.88	99.8
Italy	0.07	100.0
Jamaica	4.91	98.7
Japan	1.02	99.7

Annex 5B: Continued

Country	Diarrheal death rate	Quality indicator
Jordan	13.02	96.5
Kazakhstan	2.29	99.4
Kenya	78.06	78.9
Korea, South	0.54	99.9
Kyrgyzstan	13.04	96.5
Laos	98.48	73.4
Latvia	0.35	99.9
Lebanon	7.56	98.0
Lesotho	83.87	77.3
Liberia	150.38	59.3
Libya	8.17	97.8
Lithuana	0.12	100.0
Madagascar	111.75	69.8
Malawi	164.32	55.6
Malaysia	1.29	99.7
Maldives	16.57	95.5
Mali	180.19	51.3
Malta	0.00	100.0
Mauritania	128.48	65.3
Mauritius	1.13	99.7
Mexico	5.46	98.5
Moldova	0.76	99.8
Mongolia	33.77	90.9
Morocco	21.92	94.1
Mozambique	162.54	56.1
Namibia	43.19	88.3
Nepal	67.97	81.6
Netherlands	0.20	99.9
New Zealand	0.27	99.9
Nicaragua	26.25	92.9
Niger	219.31	40.7
Nigeria	111.21	69.9
Norway	1.98	99.5
Pakistan	78.98	78.6
Panama	6.71	98.2
Papua New Guinea	42.94	88.4

continued overleaf

Annex 5B: Continued

Country	Diarrheal death rate	Quality indicator
Paraguay	15.00	95.9
Peru	16.45	95.6
Philippines	15.56	95.8
Portugal	0.12	100.0
Romania	0.66	99.8
Russia	0.60	99.8
Rwanda	160.39	56.6
Saudi Arabia	7.16	98.1
Senegal	74.56	79.8
Sierra Leone	270.82	26.8
South Africa	30.37	91.8
Spain	0.68	99.8
Sri Lanka	3.59	99.0
Sudan	59.16	84.0
Suriname	9.43	97.4
Swaziland	57.36	84.5
Sweden	0.60	99.8
Switzerland	0.37	99.9
Syria	11.81	96.8
Tajikistan	32.80	91.1
Tanzania	87.91	76.2
Thailand	7.47	98.0
Togo	61.72	83.3
Trinidad and Tobago	1.28	99.7
Tunisia	6.75	98.2
Turkey	9.87	97.3
Uganda	120.86	67.3
Ukraine	0.32	99.9
United Kingdom	1.08	99.7
United States	0.51	99.9
Uruguay	2.80	99.2
Uzbekistan	2.12	99.4
Venezuela	6.28	98.3
Vietnam	13.39	96.4
Yemen	98.69	73.3
Zambia	144.88	60.8

Notes: Minimum diarrheal death rate: 0 (Malta)
Maximum diarrheal death rate: 369.90 (Angola)

Annex 6: IDWA-G

Country	Resource	Access	Capacity	Use	Quality	IDWA-G
Albania	58.4	70.9	58.7	88.2	100.0	75.2
Algeria	26.3	58.8	60.1	59.3	93.0	59.5
Angola	58.7	12.8	53.2	0.0	0.0	24.9
Antigua and Barbuda	32.4	82.4	78.7	62.0	99.8	71.1
Argentina	56.4	68.0	72.4	86.1	99.7	76.5
Armenia	48.2	87.7	55.1	99.3	99.5	77.9
Australia	68.6	100.0	93.4	92.2	100.0	90.8
Austria	56.0	100.0	93.7	78.6	100.0	85.6
Azerbaijan	36.7	47.0	59.9	81.6	97.0	64.4
Bangladesh	33.9	7.3	27.3	0.0	87.2	31.1
Belarus	50.6	72.3	68.4	70.5	99.9	72.3
Belgium	38.6	99.7	92.3	80.6	99.8	82.2
Belize	77.3	81.9	64.2	58.9	96.9	75.8
Benin	38.6	20.1	29.0	0.0	69.7	31.5
Bhutan	74.2	23.0	51.3	0.0	83.7	46.4
Bolivia	71.5	67.1	50.9	0.0	87.0	55.3
Botswana	40.3	52.8	73.2	60.6	94.5	64.3
Brazil	70.3	79.6	66.8	72.0	97.3	77.2
Bulgaria	47.2	89.9	69.4	55.1	99.9	72.3
Burkina Faso	36.0	2.4	26.8	0.0	53.3	23.7
Burundi	39.3	7.8	3.3	0.0	61.9	22.5
Cambodia	58.1	5.4	33.8	0.0	76.8	34.8
Cameroon	64.8	15.4	38.7	0.0	75.3	38.8
Canada	81.4	88.2	94.0	100.0	99.8	92.7
Cape Verde	31.6	48.8	43.6	0.0	93.9	43.6
Cote d'Ivoire	51.2	5.8	34.2	0.0	72.5	32.7
Central African Republic	72.1	3.0	17.3	0.0	66.8	31.9
Chad	40.9	2.7	32.0	0.0	63.1	27.8
Chile	76.6	88.5	74.0	78.1	99.6	83.4
China	44.6	68.8	54.1	48.8	97.7	62.8
Colombia	74.9	78.4	60.2	85.1	98.4	79.4
Comoros	40.7	28.6	27.1	0.0	85.9	36.5
Congo	76.4	24.7	48.6	0.0	88.3	47.6
Congo, Dem. Rep. of	63.8	17.5	0.0	0.0	40.8	24.4
Costa Rica	69.1	95.8	68.1	93.0	99.2	85.0

continued overleaf

Annex 6: Continued

Country	Resource	Access	Capacity	Use	Quality	IDWA-G
Cyprus	36.3	100.0	87.3	77.9	99.9	80.3
Czech Republic	39.7	95.4	84.2	81.0	100.0	80.1
Denmark	38.1	100.0	93.5	74.6	99.6	81.2
Djibouti	27.2	27.1	37.5	91.9	67.4	50.2
Dominican Republic	45.4	70.4	58.6	85.7	96.1	71.2
Ecuador	71.4	64.1	62.4	89.9	95.8	76.7
Egypt	0.0	84.4	55.4	76.0	95.0	62.1
El Salvador	46.3	64.9	58.3	64.3	96.0	65.9
Equatorial Guinea	75.8	6.0	88.2	93.2	74.4	67.5
Eritrea	32.0	8.9	17.1	0.0	82.5	28.1
Estonia	59.6	85.3	81.3	71.3	100.0	79.5
Ethiopia	41.2	15.0	15.8	0.0	75.2	29.5
Fiji	71.9	19.8	53.7	0.0	98.2	48.7
Finland	67.0	96.0	92.0	71.6	99.8	85.3
France	47.8	98.9	91.4	81.9	99.7	83.9
Gabon	84.2	45.2	75.7	61.7	91.4	71.6
Gambia	43.5	40.5	26.8	0.0	81.7	38.5
Georgia	62.8	62.4	51.3	88.5	99.8	73.0
Germany	39.7	99.2	91.6	72.1	99.9	80.5
Ghana	39.9	19.6	28.7	0.0	87.2	35.1
Greece	53.5	98.4	91.0	82.8	100.0	85.1
Guatemala	58.1	55.2	56.2	0.0	92.4	52.4
Guinea	68.3	9.6	27.2	0.0	69.3	34.9
Guinea-Bissau	59.5	6.0	10.3	0.0	69.7	29.1
Guyana	94.4	56.6	48.9	59.5	89.2	69.7
Haiti	41.1	20.1	28.4	0.0	81.7	34.3
Honduras	62.4	74.5	48.9	0.0	93.3	55.8
Hungary	32.1	93.3	80.6	71.7	99.9	75.5
Iceland	100.0	100.0	94.1	91.0	99.9	97.0
India	38.2	30.9	41.9	64.1	88.2	52.7
Indonesia	62.2	16.0	48.4	42.9	95.6	53.0
Iran	43.0	82.0	68.7	74.8	96.6	73.0
Ireland	61.4	97.7	95.8	72.1	100.0	85.4
Israel	15.1	99.8	85.9	81.3	99.8	76.4
Italy	48.5	98.7	89.5	86.7	100.0	84.7
Jamaica	49.7	69.9	63.5	66.0	98.7	69.6

Annex 6: Continued

Country	Resource	Access	Capacity	Use	Quality	IDWA-G
Japan	49.2	97.0	91.3	86.8	99.7	84.8
Jordan	15.7	93.2	54.1	58.3	96.5	63.5
Kazakhstan	53.3	56.9	68.6	61.4	99.4	67.9
Kenya	31.7	20.7	31.9	0.0	78.9	32.6
Korea, South	40.1	85.7	85.0	87.3	99.9	79.6
Kyrgyzstan	58.5	51.2	36.0	69.3	96.5	62.3
Laos	71.1	19.4	37.7	0.0	73.4	40.3
Latvia	56.9	81.6	77.2	71.5	99.9	77.4
Lebanon	39.9	98.1	68.4	86.1	98.0	78.1
Lesotho	47.8	10.4	31.5	0.0	77.3	33.4
Liberia	77.4	2.7	3.3	0.0	59.3	28.6
Libya	14.2	54.4	71.8	82.8	97.8	64.2
Lithuana	52.3	80.1	77.7	77.0	100.0	77.4
Madagascar	65.4	5.4	22.0	27.3	69.8	38.0
Malawi	39.0	10.4	17.6	0.0	55.6	24.5
Malaysia	67.9	94.5	73.3	71.5	99.7	81.3
Maldives	13.0	35.0	55.6	0.0	95.5	39.8
Mali	51.4	71.2	25.6	64.0	51.3	52.7
Malta	16.6	99.8	83.9	80.8	100.0	76.2
Mauritania	16.5	59.0	36.8	67.9	65.3	49.1
Mauritius	44.9	81.2	70.0	90.8	99.7	77.3
Mexico	50.2	89.7	72.7	85.9	98.5	79.4
Moldova	23.0	37.3	41.2	65.9	99.8	53.4
Mongolia	62.5	34.2	44.9	49.7	90.9	56.4
Morocco	36.1	49.7	50.8	58.7	94.1	57.9
Mozambique	52.9	6.6	18.7	0.0	56.1	26.9
Namibia	48.0	28.7	54.8	53.4	88.3	54.7
Nepal	56.5	5.1	24.5	0.0	81.6	33.5
Netherlands	33.2	100.0	93.9	37.9	99.9	73.0
New Zealand	80.6	97.5	87.0	98.0	99.9	92.6
Nicaragua	71.9	56.9	44.3	55.0	92.9	64.2
Niger	22.8	6.8	15.5	0.0	40.7	17.2
Nigeria	41.9	15.5	33.7	0.0	69.9	32.2
Norway	80.8	100.0	100.0	82.6	99.5	92.6
Pakistan	25.9	30.3	41.1	0.0	78.6	35.2
Panama	74.7	88.1	67.4	92.0	98.2	84.1

continued overleaf

Annex 6: Continued

Country	Resource	Access	Capacity	Use	Quality	IDWA-G
Papua New Guinea	85.5	11.0	36.0	0.0	88.4	44.2
Paraguay	63.7	44.1	51.4	0.0	95.9	51.0
Peru	77.1	64.7	62.3	69.7	95.6	73.9
Philippines	54.2	35.2	46.6	70.4	95.8	60.5
Portugal	43.9	97.8	83.0	82.3	100.0	81.4
Romania	49.9	50.9	69.7	79.2	99.8	69.9
Russia	71.0	73.7	74.2	80.3	99.8	79.8
Rwanda	37.3	13.1	18.6	0.0	56.6	25.1
Saudi Arabia	13.7	77.0	84.4	75.9	98.1	69.8
Senegal	44.6	43.9	33.4	0.0	79.8	40.3
Sierra Leone	70.1	11.7	15.6	0.0	26.8	24.8
South Africa	36.6	70.6	67.1	77.0	91.8	68.6
Spain	46.5	99.3	89.2	83.0	99.8	83.6
Sri Lanka	45.8	21.1	50.0	0.0	99.0	43.2
Sudan	34.9	38.0	37.2	47.7	84.0	48.4
Suriname	89.4	76.7	64.4	69.6	97.4	79.5
Swaziland	46.6	24.0	54.2	0.0	84.5	41.9
Sweden	66.3	100.0	92.6	84.5	99.8	88.6
Switzerland	54.2	99.7	94.3	77.3	99.9	85.1
Syria	26.8	79.6	52.3	50.3	96.8	61.2
Tajikistan	59.9	40.6	33.7	75.1	91.1	60.1
Tanzania	44.6	5.3	24.4	0.0	76.2	30.1
Thailand	48.8	28.6	63.6	31.0	98.0	54.0
Togo	42.9	8.2	19.6	0.0	83.3	30.8
Trinidad and Tobago	47.8	68.7	80.0	89.7	99.7	77.2
Tunisia	28.3	75.9	61.6	54.3	98.2	63.7
Turkey	48.1	88.2	65.6	76.4	97.3	75.1
Uganda	39.5	3.0	22.3	0.0	67.3	26.4
Ukraine	38.7	72.8	59.7	78.6	99.9	70.0
United Kingdom	45.9	99.2	92.0	50.7	99.7	77.5
United States	59.2	86.3	97.5	94.8	99.9	87.5
Uruguay	65.2	96.2	69.3	30.8	99.2	72.1
Uzbekistan	32.1	48.4	39.6	83.7	99.4	60.7
Venezuela	69.5	83.6	70.9	89.0	98.3	82.3
Vietnam	51.5	26.7	41.1	73.5	96.4	57.8
Yemen	20.4	26.2	40.3	0.0	73.3	32.0
Zambia	56.0	18.7	28.9	27.2	60.8	38.3

Annex 7: IDWA-G and HDI, WGI and CPI

Country	IDWA-G	HDI	WGI	CPI
Albania	75.2	0.82	45.0	3.2
Algeria	59.5	0.75	26.1	2.8
Angola	24.9	0.56	15.2	1.9
Antigua and Barbuda	71.1	0.87	74.2	N/A
Argentina	76.5	0.87	41.4	2.9
Armenia	77.9	0.80	44.1	2.7
Australia	90.8	0.97	94.1	8.7
Austria	85.6	0.96	95.0	7.9
Azerbaijan	64.4	0.79	25.8	2.3
Bangladesh	31.1	0.54	20.2	2.4
Belarus	72.3	0.83	21.7	2.4
Belgium	82.2	0.95	87.3	7.1
Belize	75.8	0.77	50.0	N/A
Benin	31.5	0.49	43.7	2.9
Bhutan	46.4	0.62	52.9	5.0
Bolivia	55.3	0.73	24.6	2.7
Botswana	64.3	0.69	72.1	5.6
Brazil	77.2	0.81	52.7	3.7
Bulgaria	72.3	0.84	59.9	3.8
Burkina Faso	23.7	0.39	38.9	3.6
Burundi	22.5	0.39	15.1	1.8
Cambodia	34.8	0.59	22.1	2.0
Cameroon	38.8	0.52	21.3	2.2
Canada	92.7	0.97	93.8	8.7
Cape Verde	43.6	0.71	67.2	5.1
Cote d'Ivoire	32.7	0.48	8.8	2.1
Central African Republic	31.9	0.37	10.4	2.0
Chad	27.8	0.39	5.3	1.6
Chile	83.4	0.88	82.6	6.7
China	62.8	0.77	39.1	3.6
Colombia	79.4	0.81	42.5	3.7
Comoros	36.5	0.58	16.1	2.3
Congo	47.6	0.60	13.0	1.9
Congo, Dem. Rep. of	24.4	0.39	4.1	1.9
Costa Rica	85.0	0.85	67.8	5.3
Cyprus	80.3	0.91	80.3	4.1

continued overleaf

Annex 7: Continued

Country	IDWA-G	HDI	WGI	CPI
Czech Republic	80.1	0.90	78.2	4.9
Denmark	81.2	0.96	96.1	9.3
Djibouti	50.2	0.52	28.5	2.8
Dominican Republic	71.2	0.78	42.3	3.0
Ecuador	76.7	0.81	20.3	2.2
Egypt	62.1	0.70	35.3	2.8
El Salvador	65.9	0.75	48.4	3.4
Equatorial Guinea	67.5	0.72	10.4	1.8
Eritrea	28.1	0.47	13.2	2.6
Estonia	79.5	0.88	81.6	6.6
Ethiopia	29.5	0.41	23.4	2.7
Fiji	48.7	0.74	32.2	N/A
Finland	85.3	0.96	97.3	8.9
France	83.9	0.96	86.0	6.9
Gabon	71.6	0.76	28.9	2.9
Gambia	38.5	0.46	33.9	2.9
Georgia	73.0	0.78	46.9	4.1
Germany	80.5	0.95	91.6	8.0
Ghana	35.1	0.53	53.6	3.9
Greece	85.1	0.94	68.3	3.8
Guatemala	52.4	0.70	32.1	3.4
Guinea	34.9	0.44	6.8	1.8
Guinea-Bissau	29.1	0.40	15.0	1.9
Guyana	69.7	0.73	37.7	2.6
Haiti	34.3	0.53	13.3	1.8
Honduras	55.8	0.73	31.7	2.5
Hungary	75.5	0.88	75.9	5.1
Iceland	97.0	0.97	93.6	8.7
India	52.7	0.61	46.1	3.4
Indonesia	53.0	0.73	35.5	2.8
Iran	73.0	0.78	16.9	1.8
Ireland	85.4	0.97	93.5	8.0
Israel	76.4	0.94	68.4	6.1
Italy	84.7	0.95	67.7	4.3
Jamaica	69.6	0.77	49.6	3.0

Annex 7: Continued

Country	IDWA-G	HDI	WGI	CPI
Japan	84.8	0.96	84.3	7.7
Jordan	63.5	0.77	53.1	5
Kazakhstan	67.9	0.80	33.6	2.7
Kenya	32.6	0.54	28.2	2.2
Korea, South	79.6	0.94	71.3	5.5
Kyrgyzstan	62.3	0.71	23.1	1.9
Laos	40.3	0.62	17.1	2.0
Latvia	77.4	0.87	69.9	4.5
Lebanon	78.1	0.80	27.5	2.5
Lesotho	33.4	0.51	45.6	3.3
Liberia	28.6	0.44	18.9	3.1
Libya	64.2	0.85	25.4	2.5
Lithuana	77.4	0.87	71.9	4.9
Madagascar	38.0	0.54	40.6	3.0
Malawi	24.5	0.49	39.8	3.3
Malaysia	81.3	0.83	58.9	4.5
Maldives	39.8	0.77	39.7	2.5
Mali	52.7	0.37	39.2	2.8
Malta	76.2	0.90	88.0	5.2
Mauritania	49.1	0.52	20.6	2.5
Mauritius	77.3	0.80	75.1	5.4
Mexico	79.4	0.85	46.7	3.3
Moldova	53.4	0.72	35.6	3.3
Mongolia	56.4	0.73	41.8	2.7
Morocco	57.9	0.65	43.6	3.3
Mozambique	26.9	0.40	40.5	2.5
Namibia	54.7	0.69	66.6	4.5
Nepal	33.5	0.55	22.8	2.3
Netherlands	73.0	0.96	94.0	8.9
New Zealand	92.6	0.95	95.5	9.4
Nicaragua	64.2	0.70	28.9	2.5
Niger	17.2	0.34	25.2	2.9
Nigeria	32.2	0.51	17.7	2.5
Norway	92.6	0.97	96.3	8.6
Pakistan	35.2	0.57	20.8	2.4
Panama	84.1	0.84	58.2	3.4

continued overleaf

Annex 7: Continued

Country	IDWA-G	HDI	WGI	CPI
Papua New Guinea	44.2	0.54	26.1	2.1
Paraguay	51.0	0.76	24.8	2.1
Peru	73.9	0.81	42.0	3.7
Philippines	60.5	0.75	37.3	2.4
Portugal	81.4	0.91	84.1	5.8
Romania	69.9	0.84	57.2	3.8
Russia	79.8	0.82	26.2	2.2
Rwanda	25.1	0.46	38.2	3.3
Saudi Arabia	69.8	0.84	45.1	4.3
Senegal	40.3	0.46	43.2	3.0
Sierra Leone	24.8	0.37	22.1	2.2
South Africa	68.6	0.68	62.9	4.7
Spain	83.6	0.96	77.8	6.1
Sri Lanka	43.2	0.76	39.4	3.1
Sudan	48.4	0.53	4.2	1.5
Suriname	79.5	0.77	49.2	3.7
Swaziland	41.9	0.57	34.5	3.6
Sweden	88.6	0.96	96.2	9.2
Switzerland	85.1	0.96	96.4	9.0
Syria	61.2	0.74	19.6	2.6
Tajikistan	60.1	0.69	15.1	2.0
Tanzania	30.1	0.53	41.9	2.6
Thailand	54.0	0.78	43.4	3.4
Togo	30.8	0.50	18.8	2.8
Trinidad and Tobago	77.2	0.84	57.6	3.6
Tunisia	63.7	0.77	50.4	4.2
Turkey	75.1	0.81	50.1	4.4
Uganda	26.4	0.51	33.0	2.5
Ukraine	70.0	0.80	37.0	2.2
United Kingdom	77.5	0.95	89.2	7.7
United States	87.5	0.96	87.3	7.5
Uruguay	72.1	0.87	71.5	6.7
Uzbekistan	60.7	0.71	12.3	1.7
Venezuela	82.3	0.84	12.7	1.9
Vietnam	57.8	0.73	34.6	2.7
Yemen	32.0	0.58	17.2	2.1
Zambia	38.3	0.48	41.0	3.0

4

Among the Indian States

Seetharam Kallidaikurichi E. and Bhanoji Rao

*IDWA was first introduced in the Asian Water and Development
Outlook, 2007, a report issued by the Asian Development Bank.
The computation of the index (more recently christened IDWA-I)
calls for data on 5 different parameters: resources, capacity,
access, quality and use. For 28 Indian states, we could manage
to obtain data on 4 parameters (resources, access, capacity and
quality) and for 15 states on 'use' also. All data refer to 2001 or
years close to that. Access, however, has two variants: general
access that refers to access to water via taps, hand pumps and tube
wells, and 'optimal access' via taps within residential premises.
The two variants provide the basis for computing IDWA-I and
IDWA-II. Inter-relationships between the two and between them
and a couple of development indicators are also explored and
implications noted. It is gratifying to note from the inter-correlations
that there is really no second best to a tap in house when it comes
to human development.*

1. Introduction

The National Water Policy 2002 has assigned the highest priority
for drinking water supply needs followed by irrigation, hydro-power,
navigation and industrial and other uses. The policy envisages the

provision for drinking water is made in all water resources projects. The drinking water requirements of most of the mega cities/cities in India are met from reservoirs of irrigation/multi-purpose schemes existing in near by areas and even by long distance transfer. Thus, Delhi gets drinking water from Tehri Dam and Chennai city from Krishna Water through Telugu Ganga Project.

Despite various laudable initiatives and programs, it is abundantly clear from the onslaught of diseases such as diarrhoea that water inadequacies continue to plague the people of the country, and one must expect serious differences in water adequacy across the various states.

This paper aims to articulate the inter-state differences in household water adequacy by recourse to the index of drinking water adequacy. Rest of the paper is organized as follows. The making of IDWA-I and IDWA-II for the Indian states is explained in the next section. In Section 3, we explore inter-relationships between them as well as between each of them and poverty and human development indicators. Section 4 has concluding observations.

2.   IDWA-I and IDWA-II Estimates

The Index of Drinking Water Adequacy (since known as IDWA-I) was first conceived and computed for 23 ADB member countries as part of the preparatory work for the maiden issue of the Asian Water and Development Outlook, 2007, brought out by the Asian Development Bank. More recently Seetharam and Rao (2010) have provided IDWA-I and IDWA-II for the same 23 countries.

To compile IDWA for Indian states, on a basis similar to the earlier estimates for the Asian economies, data are required on the following parameters: Resources, Access, Capacity, Quality and Use. The data and estimates are explained in detail below.

Resources

Ideally we should be taking the net annual recharge of both ground water and surface water on a per capita basis as the indicator of

Table 1: Projected Water Demand for Different Uses in India, 1990–2050

(Unit : Billion Cubic Meters)

Sector	1990	2000	2010	2025	2050
Domestic	32	42	56	73	102
Drinking (incl. live stock)	–	–	56	73	–
Irrigation	437	541	688	910	1,072
Industry	–	8	12	23	63
Energy	–	2	5	15	130
Other	33	41	52	72	80
Total	502	634	813	1,093	1,447

Source: Water and Related Statistics, March 2002, Central Water Commission. [Based on Indian Parliament (Lok Sabha) Question No. 2902, dated 03.09.2007.]

resource. In the earlier investigations, 'resources' referred to renewable internal fresh water resources per capita. Such data are available only at the national level and not sub-regional level. Renewable groundwater resources, however, are available at the state level. The data suitably adjusted on a per capita basis could be used as an indicator of state level water resources, provided we have some norm against which to judge the resources in each state. Table 1 helps in arriving at the norm as follows. The average water demand over the two years 2000 and 2010 is 723.5 BCM. Average population is 1.0727. Per capita demand is 674.5 BCM, or equivalent LPCD of 1848. Table 2 has the resource index. Where the index is above 100, we simply peg it at 100 since the idea is to explore adequacy and not over and above adequacy.

Access

Access is measured as the percent of population with access to a sustainable 'improved' water source. As seen in Table 3, improved access comprises water available via tap, hand pump and tube well. The per cent of population with access from any of the three is

Table 2: Resource Index for Indian States

State	BCM/Yr*	2001–4 Av Pop 000	Lt/cap/p.a.	Lt/cap/day	Index % of 1848
Andhra Pradesh	35.29	77,127.5	457,554	1,253.6	67.8
Arunachal Pradesh	1.44	1,115	1,291,480	3,538.3	100
Assam	22.48	27,344	822,118	2,252.4	100
Bihar	26.98	85,344.5	316,131	866.1	46.9
Delhi	0.29	14,455.5	20,062	55.0	3.0
Goa	0.22	1,397.5	157,424	431.3	23.3
Gujarat	20.38	51,896	392,708	1,075.9	58.2
Haryana	11.18	21,689.5	515,457	1,412.2	76.4
Himachal Pr.	0.29	6,185.5	46,884	128.4	7.0
Jammu and Kashmir	4.42	10,502.5	420,852	1,153.0	62.4
Jharkhand	6.6	27,606	239,078	655.0	35.4
Karnataka	16.17	53,779	300,675	823.8	44.6
Kerala	7.9	32,350.5	244,200	669.0	36.2
Madhya Pradesh	34.82	62,311	558,810	1,531.0	82.8
Maharashtra	37.87	99,013.5	382,473	1,047.9	56.7
Manipur	3.15	2,444	1,288,871	3,531.2	100
Meghalaya	0.54	2,358.5	228,959	627.3	33.9
Mizoram	1.4	911.5	1,535,930	4,208.0	100
Nagaland	0.72	2,039.5	353,028	967.2	52.3
Orissa	20.13	37,423	537,904	1,473.7	79.7
Punjab	18.19	24,812.5	733,098	2,008.5	100
Rajasthan	12.6	58,300	216,123	592.1	32.0
Sikkim	0.07	553	126,582	346.8	18.8
Tamil Nadu	26.41	63,065	418,774	1,147.3	62.1
Tripura	0.66	3,258.5	202,547	554.9	30.0
Uttar Pradesh	82.55	171,409	481,597	1,319.4	71.4
Uttaranchal	2.84	8,702.5	326,343	894.1	48.4
West Bengal	23.09	81,903	281,919	772.4	41.8

Note: *The ground water resources of the country have been estimated for freshwater based on the guidelines and recommendations of the Ground Water Estimation Commission 1997. The total annual replenishable ground water resources of the country have been estimated as 433 billion cubic meter (BCM). Keeping 34 BCM for natural discharge, the net annual ground water availability for the entire country is 399 BCM. The annual ground water draft is 231 BCM out of which 213 BCM is for irrigation use and 18 BCM is for domestic and industrial use.

Source of basic data: Central Ground Water Board.

Table 3: Percent of Population with Access to 'Safe' Drinking Water, 2001 Census Data

State	Tap	Hand pump	Tube well	Total: Access Indicator
Andhra Pradesh	48.1	26.1	5.9	80.1
Arunachal Pradesh	67.8	7.4	2.4	77.6
Assam	9.2	44.6	5	58.8
Bihar	3.7	77.9	5	86.6
Delhi	75.3	18.7	3.2	97.2
Goa	69	0.6	0.4	70.0
Gujarat	62.3	16.7	5.1	84.1
Haryana	48.1	31.7	6.2	86.0
Himachal Pradesh	84.1	4	0.5	88.6
Jammu and Kashmir	52.5	11.3	1.4	65.2
Jharkhand	12.6	27	3.1	42.7
Karnataka	58.9	17.1	8.6	84.6
Kerala	20.4	1.1	1.9	23.4
Madhya Pradesh	25.3	39.2	3.9	68.4
Maharashtra	64	12.9	2.9	79.8
Manipur	29.3	6.4	1.3	37.0
Meghalaya	34.5	2	2.5	39.0
Mizoram	31.9	1.9	2.1	35.9
Nagaland	42	2.5	2	46.5
Orissa	8.7	28.5	27	64.2
Punjab	33.6	60.8	3.2	97.6
Rajasthan	35.3	26.4	6.6	68.3
Sikkim	70.3	0.2	0.2	70.7
Tamil Nadu	62.5	17.8	5.2	85.5
Tripura	24.6	14.9	13.1	52.6
Uttar Pradesh	23.7	63.4	0.7	87.8
Uttaranchal	65.9	19.8	1	86.7
West Bengal	21.4	55.8	11.3	88.5

totaled to arrive at overall access. The data are from the Indian Population Census of 2001.

In IDWA-I, we use general access to water in line with the data in the last column of Table 3, and in tune with the water access goals in the Millennium Development Goals, which refer to "access to a sustainable improved water source". Such access, however, is considered sub-optimal in terms of ensuring minimal health risks (Seetharam and Rao, 2010). Indeed, if one were to factor in not only health risk, but also the opportunity cost of time lost in collecting water and related factor, the intrinsic merits of water connection at home becomes indisputably clear. Hence, in IDWA-II, access via home connection is used in place of general access. [The inter-correlation between rates of general access and access via tap at home was only 0.5.]

Capacity

Per capita GDP is used as a general proxy for the capacity to purchase water. Table 4 has the details on the data as well as the way the capacity index is obtained. In line with the data on resources, an attempt is made to position the per capita GDP for the year 2003, but, to minimize the effect, if any, of year to year transitory varia-tions, three year average is used. As explained in the note under the Table, Goa with the highest per capita GDP and Bihar with the lowest serve as the benchmarks in preparing the capacity index.

Quality

India has fairly well articulated standards for monitoring water quality.[1] Despite the standards and the availability of considerable data on quality monitoring of rivers and ground water bodies, state-wise data on the quality for drinking water are not available.[2] We, therefore, follow the earlier work and make use of one major disease that reflects water quality: diarrhoea. The quality index

---

[1] See Bhardwaj (2005) for an articulation of the detailed standards.

[2] See, however, Khurana and Sen (2008) for a listing of states under different types of quality parameters like fluoride and arsenic contamination and other pertinent qualitative information. The usual data on access to water routinely published by national and international organizations do not quite pay attention to quality of water. For an exposition of the problem, see Lodhia (2006).

Table 4: Per Capita Gross Domestic Product and Index of Capacity to Purchase Water

State	Per capita GDP					
	2002–3	2003–4	2004–5	Average	Log of Av	Index
Andhra Pradesh	19,568	22,041	23,729	21,779	4.34	54.0
Arunachal Pradesh	17,434	19,707	22,542	19,894	4.30	49.7
Assam	14,600	15,687	16,825	15,704	4.20	38.3
Bihar	6,928	6,913	7,467	7,103	3.85	0.0
Delhi	45,483	49,825	55,215	50,174	4.70	94.3
Goa	48,839	54,577	66,135	56,517	4.75	100.0
Gujarat	22,683	26,922	29,468	26,358	4.42	63.2
Haryana	28,259	31,509	35,044	31,604	4.50	72.0
Himachal Pradesh	26,627	28,333	31,140	28,700	4.46	67.3
Jammu and Kashmir	16,452	17,528	18,630	17,537	4.24	43.6
Jharkhand	11,865	12,941	17,493	14,100	4.15	33.1
Karnataka	19,041	20,515	24,199	21,252	4.33	52.8
Kerala	23,207	25,645	27,864	25,572	4.41	61.8
Madhya Pradesh	12,303	14,306	14,534	13,714	4.14	31.7
Maharashtra	26,697	29,770	32,979	29,815	4.47	69.2
Manipur	13,250	14,728	18,386	15,455	4.19	37.5
Meghalaya	18,756	20,729	21,915	20,467	4.31	51.0
Mizoram	20,896	21,963	22,417	21,759	4.34	54.0
Nagaland	20,407	20,821	20,998	20,742	4.32	51.7
Orissa	11,788	14,252	16,306	14,115	4.15	33.1
Pondicherry	44,903	48,547	44,908	46,119	4.66	90.2
Punjab	29,443	31,192	32,945	31,193	4.49	71.3
Rajasthan	13,126	16,704	16,800	15,543	4.19	37.8
Sikkim	19,428	21,476	23,791	21,565	4.33	53.5
Tamil Nadu	21,813	24,106	27,137	24,352	4.39	59.4
Tripura	19,059	21,138	22,836	21,011	4.32	52.3
Uttar Pradesh	10,435	11,250	11,941	11,209	4.05	22.0
Uttaranchal	18,891	20,519	22,093	20,501	4.31	51.1
West Bengal	18,746	20,806	22,522	20,691	4.32	51.6

Note: Index in the last column is based on taking the values of zero and 100 respectively for Bihar and Goa. Bihar's per capita GDP of Rs. 7103 translates to a little less than Rs. 600 per capita per month, too meager to denote any capacity to purchase water. Goa, on the other hand has an annual per capita GDP of Rs. 56517, or about Rs. 4710 per month.

Source of basic data: India, Economic Survey, 2008.

Table 5:   Quality Index (Proxy: Diarrhoea death rate)

	Diarrhoea Deaths			Population 2006 (000)	Deaths per million people	Index
	2006	*2007*	*Average*			
Andhra Pradesh	124	198	161	80,430	2.00	98.0
Arunachal Pradesh	30		30	1,170	25.64	74.4
Assam		911	911	29,009	31.40	68.6
Bihar				90,830	**1.31**	98.7
Delhi	85	70	77.5	16,065	4.86	95.1
Goa		0	0	1,537	0.00	100.0
Gujarat	4	3	3.5	54,814	0.07	99.9
Haryana	42	30	36	23,040	1.56	98.4
Himachal Pr.	28	33	30.5	6,425	4.82	95.2
Jammu and Kashmir	32		32	11,603	2.76	97.2
Jharkhand	1	6	3.5	29,173	0.14	99.9
Karnataka	1,279	80	679.5	56,137	12.11	87.9
Kerala	4	12	8	33,569	0.24	99.8
Madhya Pradesh	88	302	195	66,801	2.92	97.1
Maharashtra	93	199	146	104,104	1.40	98.6
Manipur	17	16	16.5	2,561	6.64	93.4
Meghalaya	33	60	46.5	2,472	19.01	81.0
Mizoram	20	10	15	955	15.71	84.3
Nagaland	0	6	3	2,132	1.41	98.6
Orissa	40	68	54	39,053	1.38	98.6
Punjab	64	84	74	25,976	2.85	97.2
Rajasthan	21	38	29.5	62,431	0.48	99.5
Sikkim	8	9	8.5	579	15.54	84.5
Tamil Nadu	42	140	91	65,261	1.39	98.6
Tripura	69	19	44	3,421	12.86	87.1
Uttar Pradesh	55	137	96	183,856	0.52	99.5
Uttaranchal	6	18	12	9,216	1.30	98.7
West Bengal	964	1,118	1,041	85,780	12.14	87.9

Note: Estimate for Bihar is based on the average of Madhya Pradesh, Rajasthan and Uttar Pradesh.

based on diarrhoeal deaths (per million people) is shown in Table 5. It may be noted that by simply taking the zero death rate of Goa as the norm, and equating it to 100, rest of the indicators are worked out.

Table 6: Average Water Use by Households in Selected Cities

State	City	LPCD	City	LPCD	City	LPCD	Average LPCD
Andhra Pradesh	Hyderabad	90	Visakha	58	Vijawada	137	95
Bihar	Patna	107					107
Delhi	Delhi	155					155
Gujarat	Ahmadabad	116	Surat	180	Vadodara	135	130
	Rajkot	88					
Haryana	Faridabad	120					120
Jharkhand	Jamshedpur	90					90
Karnataka	Bangalore	100					100
Kerala	Kochi	129					129
Madhya Pradesh	Indore	80	Bhopal	130	Jabalpur	95	152
Maharashtra	Mumbai	170	Pune	220	Nagpur	130	
	Nashik	140					165
Punjab	Ludhiana	140	Amritsar	135			137
Rajasthan	Jaipur	97					97
Tamil Nadu	Chennai	65	Coimbatore	109	Madurai	72	82
Uttar Pradesh	Kanpur	118	Lucknow	264	Agra	134	
	Varanasi	103	Meerut	185	Allahabad	111	152
West Bengal	Kolkata	100	Asansol	120	Dhanbad	70	97

Source of basic data: Indian Water Utilities Databook, 2003.

Use

It is one thing to have supply of some amount of drinking water. Adequacy in terms of quantity and quality are both important. In earlier papers, since they dealt with the national level, it was relatively easy to assemble the per capita use of water for domestic purposes. Such data are not available at the sub-national level in India. A number of utilities, however, provide the data for selected cities in each state. These are shown in Table 6. The last column of the table shows the average LPCD taken as the best possible estimate of use, which applies to the urban sector.

As can be seen from the Table, for only 15 of the 28 states, we could get some data on water use in urban areas. In line with the urban character of the data, the norms used to judge adequacy are based on the best and the minimum. For best, the norm refers to the average per capita domestic consumption over 1995–2002 in Singapore (167 LPCD), where water conservation is combined with guaranteed 24×7 supply. For the minimum, we use the norm of 70 LPCD suggested by Government of India.[3] Based on the norms of 167 LPCD and 70 LPCD, and the average use in Table 6, the index of use is computed as follows. Taking the average use U_j observed for state j, we have

$$\text{Use Indicator for 'j'} = \left[\frac{(U_j - 70)}{(167 - 70)} \right] \times 100$$

The data on the computed index for the 15 states is given in Table 7 below. It is pertinent to note that the highest indicator values for use are noted in respect of urban Maharashtra and urban Delhi. A surprise finding is the extremely low level recorded for Tamil Nadu.

---

[3] Under the Indian Government's Accelerated Urban Water Supply Program taken up in the Eighth Five Year Plan (1992–97), the norm for drinking water was set at 70 LPCD. Other norms were given as follows: with sewerage: 125 LPCD; without sewerage: 70 LPCD; and with spot sources and public stand posts: 40 LPCD.

Table 7: Index of Urban Domestic Water Use for 15 States

Andhra Pradesh	25.8	Jharkhand	20.6	Punjab	69.1
Bihar	38.1	Karnataka	30.9	Rajasthan	27.8
Delhi	87.6	Kerala	60.8	Tamil Nadu	12.4
Gujarat	61.9	Madhya Pradesh	84.5	Uttar Pradesh	84.5
Haryana	51.5	Maharashtra	97.9	West Bengal	27.8

IDWA-I and IDWA-II across the States

IDWA-I (four/five components), and IDWA-II based on house connections for 15/28 states are shown in Tables 8, 9 and 10. Based on Table 8, the correlation coefficient between the two was found to be 0.89, high but not perfect. We should in general prefer the index based on the five components though one must manage with four components due to data limitations.

Table 8: IDWA-I for 15 States with 4 and 5 Components

	Capacity	Access	Resource	Quality	Use	IDWA-I [4 com]	IDWA-I [5 com]
Punjab	71.3	97.6	100	97.2	69.1	91.5	87.0
Maharashtra	69.2	79.8	56.7	98.6	97.9	76.1	80.4
Haryana	72.0	86.0	76.4	98.4	51.5	83.2	76.9
Delhi	94.3	97.2	3.0	95.1	87.6	72.4	75.4
Gujarat	63.2	84.1	58.2	99.9	61.9	76.4	73.5
Uttar Pradesh	22.0	87.8	71.4	99.5	84.5	70.2	73.0
Madhya Pradesh	31.7	68.4	82.8	97.1	84.5	70.0	72.9
Andhra Pradesh	54.0	80.1	67.8	98.0	25.8	75.0	65.1
Tamil Nadu	59.4	85.5	62.1	98.6	12.4	76.4	63.6
Karnataka	52.8	84.6	44.6	87.9	30.9	67.5	60.2
West Bengal	51.6	88.5	41.8	87.9	27.8	67.4	59.5
Kerala	61.8	23.4	36.2	99.8	60.8	55.3	56.4
Bihar	0.0	86.6	46.9	98.7	38.1	58.0	54.1
Rajasthan	37.8	68.3	32.0	99.5	27.8	59.4	53.1
Jharkhand	33.1	42.7	35.4	99.9	20.6	52.8	46.3

Table 9: IDWA-II for 15 States

	Capacity	*H-conn*	*Resource*	*Quality*	*Use*	*IDWA-II*
Maharashtra	69.2	45.2	56.7	98.6	97.9	73.5
Punjab	71.3	28.4	100.0	97.2	69.1	73.2
Delhi	94.3	60.5	3.0	95.1	87.6	68.1
Gujarat	63.2	42.4	58.2	99.9	61.9	65.1
Haryana	72.0	26.7	76.4	98.4	51.5	65.0
Madhya Pradesh	31.7	15.0	82.8	97.1	84.5	62.2
Uttar Pradesh	22.0	17.5	71.4	99.5	84.5	59.0
Kerala	61.8	13.0	36.2	99.8	60.8	54.3
Andhra Pradesh	54.0	20.6	67.8	98.0	25.8	53.2
Tamil Nadu	59.4	18.8	62.1	98.6	12.4	50.3
Karnataka	52.8	24.2	44.6	87.9	30.9	48.1
Rajasthan	37.8	26.7	32.0	99.5	27.8	44.8
West Bengal	51.6	10.8	41.8	87.9	27.8	44.0
Jharkhand	33.1	8.2	35.4	99.9	20.6	39.4
Bihar	0.0	2.7	46.9	98.7	38.1	37.3

3.   Selected Inter-relationships

Between IDWA-I and IDWA-II

General access to water is not the same as having a tap in one's home. Between IDWA-I and IDWA-II, the correlation was high at 0.87 in respect of 28 states and 0.94 in regard to the 15 states sample. Despite these expectedly high correlations because of the relative dominance of the other included components [respectively 3/4 for 28/15 states], one must consider optimal access as part of water adequacy, and hence one should use IDWA-II in preference to IDWA-I.

Between IDWA and Other Development Indicators

From the websites of the National Commission on Women and the Ministry of Finance [ref: Economic Surveys], we assembled *indicators* of education, life expectancy, income, human development and poverty [Table 11]. The inter-correlations between them and

Table 10: IDWA-I for 28 States [Based on 4 Components]

	Capacity	Access-I	Access-II	Resource	Quality	IDWA-I	IDWA-II
Punjab	71.3	97.6	28.4	100.0	97.2	91.5	74.2
Haryana	72.0	86.0	26.7	76.4	98.4	83.2	68.4
Goa	100.0	70.0	46.6	23.3	100.0	73.3	67.5
Maharashtra	69.2	79.8	45.2	56.7	98.6	76.1	67.4
Gujarat	63.2	84.1	42.4	58.2	99.9	76.4	65.9
Delhi	94.3	97.2	60.5	3.0	95.1	72.4	63.2
Arunachal Pradesh	49.7	77.6	28.6	100.0	74.4	75.4	63.2
Mizoram	54.0	35.9	13.3	100.0	84.3	68.5	62.9
Andhra Pradesh	54.0	80.1	20.6	67.8	98.0	75.0	60.1
Tamil Nadu	59.4	85.5	18.8	62.1	98.6	76.4	59.7
Manipur	37.5	37.0	7.8	100.0	93.4	67.0	59.7
Uttaranchal	51.1	86.7	31.6	48.4	98.7	71.2	57.5
Jammu and Kashmir	43.6	65.2	24.0	62.4	97.2	67.1	56.8
Madhya Pradesh	31.7	68.4	15.0	82.8	97.1	70.0	56.7
Nagaland	51.7	46.5	14.6	52.3	98.6	62.3	54.3
Orissa	33.1	64.2	4.9	79.7	98.6	68.9	54.1
Assam	38.3	58.8	4.2	100.0	68.6	66.4	52.8
Kerala	61.8	23.4	13.0	36.2	99.8	55.3	52.7
Uttar Pradesh	22.0	87.8	17.5	71.4	99.5	70.2	52.6
Karnataka	52.8	84.6	24.2	44.6	87.9	67.5	52.4
Sikkim	53.5	70.7	47.2	18.8	84.5	56.9	51.0
Himachal Pradesh	67.3	88.6	32.2	7.0	95.2	64.5	50.4
Rajasthan	37.8	68.3	26.7	32.0	99.5	59.4	49.0
West Bengal	51.6	88.5	10.8	41.8	87.9	67.4	48.0
Meghalaya	51.0	39.0	13.7	33.9	81.0	51.2	44.9
Tripura	52.3	52.6	7.7	30.0	87.1	55.5	44.3
Jharkhand	33.1	42.7	8.2	35.4	99.9	52.8	44.1
Bihar	0.0	86.6	2.7	46.9	98.7	58.0	37.1

Table 11: Human Development and Poverty Indicators and IDWA

States	Education Index	Life Exp. Index	Income Index	HDI	Poverty Ratio	IDWA-I	IDWA-II
Andhra Pradesh	0.539	0.672	0.513	0.575	25.7	65.1	53.2
Bihar	0.413	0.626	0.308	0.449	55	54.1	37.3
Gujarat	0.612	0.661	0.544	0.606	24.2	73.5	65.1
Haryana	0.57	0.703	0.579	0.617	25.1	76.9	65.0
Karnataka	0.607	0.687	0.531	0.608	33.2	60.2	48.1
Kerala	0.751	0.867	0.544	0.721	25.4	56.4	54.3
Madhya Pradesh	0.569	0.552	0.447	0.523	42.5	72.9	62.2
Maharashtra	0.678	0.728	0.581	0.662	36.9	80.4	73.5
Punjab	0.58	0.766	0.589	0.645	11.8	87.0	73.2
Rajasthan	0.578	0.628	0.466	0.557	27.4	53.1	44.8
Tamil Nadu	0.662	0.702	0.549	0.638	35	63.6	50.3
Uttar Pradesh	0.456	0.587	0.423	0.489	40.9	73.0	59.0
West Bengal	0.588	0.679	0.511	0.593	35.7	59.5	44.0

Table 12: Inter-correlations

Development indicator	IDWA-I	IDWA-II
Education Index	0.05	0.31
Life Expectancy	0.05	0.22
Income Index	0.51	0.62
Human Development Index	0.22	0.42
Poverty	−0.39	−0.50

IDWA are noted in Table 12. They speak eloquently about the relatively strong relationship between development indicators and IDWA-II, the index of water adequacy that includes access via house connections.

4. Concluding Notes

Importance of House Connections and Interim Measures

It is easy to comprehend the singular importance of house connections, with the pre-condition of proper housing to host the connections. This is not to imply denial of drinking water to the vast majority of rural and urban residents, till they have proper housing, which may not be possible in the short-term. In fact the pace of growth in house connections has been extremely meager. For instance, as per Census 2001, nationally, 20.8% of the people have access to water via house connections. By 2005–6, the percentage rose just to 23.5%.[4] Thus, ad-hoc measures are needed to ensure water supply to the rest of the population until house connections take off into a relatively better expansion mode, based on sound house building programs for the low income groups.[5]

In Andhra Pradesh, for instance, it was reported[6] in July that the government was contemplating two alternative routes to provide mineral (bottled) water to the villages not covered by protected water schemes: supply of water free of cost and sale of 20 litres daily to a household for a subsidised price of Rs. 2. [Note: The production of water is planned to be from RO plants operated by women's self-help groups.] How does the monthly expense of Rs. 60 compare with average income/expenditure? A recent NSSO (National Sample Survey)[7] found that in 2006–7, average monthly per capita consumer expenditure was Rs. 695 in rural India and Rs. 1,312 in urban India. Thus expense of water of Rs. 60 would amount to 8.6% and 4.6% respectively of total expenditure.

Interim solutions such as providing water via public taps are important but may not be up to the mark as implied in some of the findings of a report by Sekhar, Nair and Reddy (2005). A survey of 297 households from 4 slums of Bangalore revealed that just about 30% of the residents reported water availability through out the

---

[4] International Institute for Population Sciences (IIPS) and Macro International (2007), *National Family Health Survey (NFHS-3), 2005–06: India: Volume I*. Mumbai: IIPS.

[5] It is high time India plans to build new cities around its thousands of small and medium towns, in part to boost economic and social development, and also to arrest the unacceptable growth of metros. The idea of a thousand cities housing the nation has been briefly articulated in Bhanoji Rao (1993).

[6] See *The Hindu*, July 1, 2009.

[7] NSSO (2008), *Level of Consumption in July 2006–June 2007*, NSSO 63rd Round Report published by NSSO, October, 2008.

year from public taps. Breakdowns are common with 77% of respondents reporting breakdowns within three months. Overall dissatisfaction with public taps was a high 60%. As expected, piped water does better: 60% of households reported availability throughout the year. Relatively more are satisfied with the service. The study also found that people are willing to pay for the connections. In sum, it could be said that pending availability of decent housing, as long as there is some dwelling, it is best to try and provide a connection rather than ask people to depend on a public tap.

Roles of Local Authorities and Integrated Management

It is often pointed out that thanks to the 73rd Amendment of the Constitution, it is now possible for urban and rural local bodies to be endowed with authority and responsibility for implementation of schemes to deliver essential services such as drinking water supply, sanitation, street lighting and roads. In fact there is some evidence (De, 2009, for instance) to show how locally managed initiatives help service quality in water supply. Based on household surveys in six villages of Birbhum district of West Bengal, De compared the households' satisfaction with water supply through decentralized and centralized institutions. Two key findings are as follows. Quality of piped water was better in villages where local government participated in O&M. Contributions to maintenance expenses by households had a positive impact on quality of water supply services. Despite the relatively low management capacity of most local bodies, and possible corruption and elite capture, enunciated by Bardhan and Mookherjee (2000) among others, one cannot dispute the efficacy of local management in the broad view of the clients.[8] In this

---

[8] In addition to local governments ensuring quality, they could well be the right agents to ensure the five pronged water demand management suggested by Brooks (2006). The five prongs are reducing the quantity or quality of water required to accomplish a specific task; adjusting the nature of the task so it can be accomplished with less water or lower quality water; reducing losses in movement from source through use to disposal; shifting time of use to off-peak periods; and increasing the ability of the system to operate during droughts.

context, it should also be noted that if capacity building is a complex task at macro level (Pres, 2008), it should be even more challenging at the local level.

Then, there is the whole issue of management of the precious water resources. Most of the ground water is not stored properly, erstwhile storage locales are not protected, and river basins are a source of uncertain supplies and politically motivated inter-state disputes.[9] Yet, the need is one of integrated management of the over-exploited basins, and well articulated and understood regional and sectoral water allocation criteria.[10]

National Perspective Plan and National Water Policy

There is no dearth of good intentions and verbose documents. We have a National Perspective Plan for water; with 'inter-linking of rivers' making the rounds off and on while rivers continue to suffer from growing pollution. The National Water Policy released in 2002, for a change is only of 10 pages length, with coverage of every important issue: need for a national policy, information system, water resources planning, institutional mechanism, water allocation priorities,[11] project planning, ground water development, drinking water, irrigation, resettlement and rehabilitation, financial and physical sustainability, participatory approach to water management, private sector participation, water quality, water zoning, conservation, flood control and management, land erosion, drought-prone area development, monitoring of projects, water sharing among states, performance improvement, maintenance and modernization, safety of structures, and science, technology and training.

---

[9] Richards and Singh (2002) deal with the issue of inter-state water disputes with reference to institutions and policies.

[10] For an articulation of the management challenges in respect of the Krishna Basin, see Gaur *et al.* (2007). See also Biswas (2008), on the challenges integrated management faces.

[11] The Policy's allocation priorities are: Drinking water, Irrigation, Hydro power, Ecology, Agro and other industries, and Navigation and other uses.

References

Asian Development Bank (2007). *Asian Water and Development Outlook, 2007.* Manila: Asian Development Bank.

Bardhan, Pranab and Dilip M.(2000). "Capture and Governance at Local and National Levels", *American Economic Review*, 90(2), 135–9.

Rao, B. (1993). *India's Economic Future: Government, People and Attitudes.* Delhi: Tata McGraw-Hill Book Company.

Bhardwaj, R.M. (2005). *Water Quality Monitoring in India: Achievements and Constraints.* IWG-Env, International Work Session on Water Statistics, Vienna.

Biswas, Asit K. (2008). Integrated Water Resources Management: Is It Working? *International Journal of Water Resources Development*, 24(1), 5–22.

Brooks, D.B. (2006). An Operational Definition of Water Demand Management, *International Journal of Water Resources Development*, 22(4), 521–8.

Central Pollution Control Board (2008). *Status of Water Quality in India–2007.* New Delhi: Ministry of Environment and Forests, 2008, p. 17.

De, I. (2009). Can Decentralization Improve Rural Water Supply Services? *Economic and Political Weekly*, 44(1), January, 3, 8–71.

Gaur, A., McCornick, P.G., Turral, H., and Acharya, S (2007). Implications of Drought and Water Regulation in the Krishna Basin, India, *International Journal of Water Resources Development*, 23(4), 583–94.

Khurana, I. and Sen, R. (2008). *Drinking Water Quality in Rural India: Issues and Approaches.* Background Paper, New Delhi: Wateraid India.

Lodhia, S. (2006). *Quality of Drinking Water in India: Highly Neglected at Policy Level*, Working Paper Number 11. Ahmedabad: Centre for Development Alternatives.

Pres, A. (2008). Capacity Building: A Possible Approach to Improved Water Resources Management, *International Journal of Water Resources Development*, 24(1), 123–9.

Richards, A. and Singh, N. (2002). Inter-state Water Disputes in India: Institutions and Policies, *International Journal of Water Resources Development*, 18(4), 611–25.

Seetharam, K.E. and Rao, B. (2010). Index of Drinking Water Adequacy for the Asian Economies, *Water Policy*, 12, Supplement 1, 135–54.

Sekhar S., Nair, Meena and Reddy, Venugopal (2005), *Are They Being Served? Citizen Report Card on Public Services for the Poor in Peri-Urban Areas of Bangalore.* Bangalore: Association for Promoting Social Action and Public Affairs Centre.

5

Across the Chinese Provinces

Fan Mingxuan

In this paper, we explore the measurement of drinking water adequacy in the provinces of China, following the earlier investigations pertaining to Asian nations and Indian states. Inter-provincial differences in IDWA in China are no less pronounced than in India, a fact that underscores the importance of working for narrowing the differentials.

1. Introduction

During the past decades, the Chinese government dedicated itself to provide to its citizens increased access to drinking water. As a result, in urban areas the *water coverage rate* (see Box 1 for definitions) reached 94.7% in 2009 (National Bureau of Statistics, (NBS) 2009). In the rural areas, population facing *drinking water difficulty* declined from 40% in 2000 to 10% in 2007 (Xinhua News Agency, 2008). More recently, policy focus shifted from increasing drinking water access to guaranteeing *drinking water safety*, which, in addition to access, takes into account quality of drinking water and reliability of the sources. By the end of 2008, according to official estimates, 29% of the rural population (203 million) and 16.8% of the urban population (100 million) did not have access to safe drinking water (Ministry of Water Resources (MWR), 2009).

In 2009, MWR set a new target to ensure drinking water safety for the entire rural population by 2013 and to establish an up to date standard drinking water security system for both rural and urban areas by 2020. Under the current governance structure of water sector in China, achieving these targets would call for concerted efforts at the provincial level.

In the aforementioned context, this paper attempts to provide estimates of the Index of Drinking Water Adequacy (IDWA) for the provinces of China, with a view to providing a basis for possible monitoring and benchmarking of the performance of drinking water sector in each province. The rest of this paper is organized as follows: Section 2 briefly introduces IDWA; Section 3 reports on IDWA I and II for the provinces in China; Section 4 provides a few inter-relationships between IDWA and other development indicators as well as a simple comparison of inter-provincial variation in China and India; and Section 5 provides concluding remarks.

Box 1
Official Definitions

Water coverage rate refers to the percentage of population within the defined urban area with access to metered tap water. (*Source*: Interview with the officials from the Statistics Division, Ministry of Housing and Urban-Rural Development)

Drinking water difficulty refers to a situation where a household does not have access to water (a) within 1000 meters horizontally and 100 meters vertically within the premises or (b) for 100 continuous days or (c) drinking water contains more than 1.1 mg of fluoride per litre. (*Source*: Hainan Bureau of Water Resource, 2006)

Drinking water safety is achieved when (a) drinking water is accessible within 800 meters horizontally and 80 meters vertically from the premises; (b) enough water is available for domestic purpose as per provincial standards; (c) drinking water is from a source which has less than 5% probability of failure in supply; and (d) the water obtained is in line with the national drinking water quality standards. (*Source*: Hainan Bureau of Water Resources, 2006)

2. Methodology

IDWA was originally proposed in the *Asian Water Development Outlook* (AWDO) 2007 of the Asian Development Bank. IDWA-I was computed for 23 Asian economies by averaging five parameters, namely water resources availability, access to improved drinking water sources, capacity to buy water, water quality and water usage. Subsequently, Seetharam and Rao (2010) brought out IDWA-II by replacing "access to improved drinking water sources" with "optimal access via home connections".[1]

Data requirement for IDWA is low, with only 5 indicators involved, which makes IDWA applicable to more countries, even the ones with comparatively meagre data. IDWA is straightforward, easier to interpret for policy makers, yet comprehensive enough to direct policy focus to the vital areas. Hence, IDWA estimation could also benefit countries with large variation across provinces/ states. IDWA for the states of India, for instance, was developed by Seetharam and Rao (2009), and same basic structure is applied in this paper to compile IDWA for the provinces of China. However, due to the differences in local definition of indicators, certain changes are made to the computation methodology. Table 1 summarizes the indicators, norms and computing methods used in this paper as well as those employed in regard to IDWA for the States of India.

---

[1] As mentioned in Seetharam and Rao (2010), the access of water via improved water sources is only sub-optimal in terms of ensuring minimal health risk. It was also pointed out that the opportunity cost of time lost in collecting water makes water connection at home an even better choice.

Table 1: Summary of Methodology

		Indicator		Norm		Index Computation	
		India	China	India	China	India	China
Resource		Renewable ground water resource per capita	Renewable internal fresh water per capita	Per capita water demand in 2010 of 1848 LPCD = 100	Per capita water demand in 2010 of 1232 LPCD = 100	*Resource Index for State j =* $100 \times \dfrac{Renewable\ Ground\ Water\ per\ capita\ in\ state\ j}{1848\ LPCD}$	*Resource Index for Province j =* $100 \times \dfrac{IRWR\ per\ capita\ in\ province\ j}{1232\ LPCD}$
Access	I	Population with access via tap, hand pump and well	Urban households with access via house connections; Rural households with access via tap, deep well, shallow well and rain water	Index = 100 when 100% of the population has access to water	Index = 100 when 100% of the households have access to water	*Access Index for State j =* $100 \times \dfrac{Population\ with\ access\ in\ state\ j}{Total\ population\ in\ state\ j}$	*Access Index for Province j =* $100 \times$ (Urban access rate in province j × percentage of urban households in province j + Rural access rate in province j × percentage of rural households in province j)
	II	Population with house connections	Households with house connections				
Capacity		Per capita GDP		log (lowest provincial per capita GDP) = 0; log (highest provincial per capita GDP) = 100.		*Capacity Index for State/Province j* $= 100 \times \dfrac{\log (per\ capita\ GDP\ j) - \log (per\ capita\ GDP\ min)}{\log (per\ capita\ GDP\ max) - \log (per\ capita\ GDP\ min)}$	

Table 1: Continued

	Indicator		Norm		Index Computation	
	India	China	India	China	India	China
Quality	Diarrhoeal Death Rate	Polluted water supplied for domestic use	Index = 100 when there are no diarrhoeal deaths	Index = 100 when there is no polluted water supplied for domestic use	Quality Index for State j = $100 - $ Diarrhoeal death rate in state j	Quality Index for Province j = $100 - \dfrac{\text{Polluted water supplied in province } j \text{ for domestic use in LPCD}}{\text{Total water supplied in province } j \text{ for domestic use in LPCD}}$
Use	Water use by households in selected cities	Water for domestic use	Singapore's per capita domestic water use between 1995 and 2002 of 167 LPCD = 100; Minimum requirement of 70 LPCD = 0	Singapore's per capita domestic water use in 2007 of 157 LPCD = 100; Minimum requirement of 75 LPCD = 0	Use Index for State j = $100 \times \dfrac{\text{per capita domestic water use in state } j - 70 \text{ LPCD}}{167 \text{ LPCD} - 70 \text{ LPCD}}$	Use Index for Province j = $100 \times \dfrac{\text{per capita domestic water use in province } j - 75 \text{ LPCD}}{157 \text{ LPCD} - 75 \text{ LPCD}}$

Source: For Indian States, see Rao (2009). For Chinese Provinces, see the discussion in Section 3.

3.   IDWA-I and IDWA-II for Provinces of China

This section discusses the computation of each component of IDWA covering the definition of indicators, data sources, proxy measures where needed, norms and results. The computed IDWA-I and IDWA-II are reported at the end of this section.

Resources

Internal renewable water resource (IRWR) per capita is used as an indicator for water resource availability. Data collection in this regard is the responsibility of MWR (Box 2). The *China Statistical Yearbook* reports IRWR, under the name of Total Amount of Water Resource, as well as IRWR per capita for each of the provinces for the years 2005–7 (NBS, 2006, 2007 and 2008a). The reported value of IRWR is cubic meter per capita which is then converted to litre per capita per day (LPCD).

Box 2
Water Resource Data

Ministry of Water Resources is the main agency to collect water related data in China. *Code of Practice for Water Resources* provides the regulated methods of collection and reporting of water related data.

Precipitation, evaporation, natural runoff, ground water recharge and discharge, etc., are recorded at hydrological stations. Based on the data collected, total amount of water resources is calculated by using the total amount of surface runoff and groundwater recharge, adjusting for duplications in measurement.

Consolidated data are reported annually at the provincial level and released in *China Water Resource Bulletin*.

Source: Code of Practice for Water Resources Bulletin.

To compute the resource index for IDWA, the available water resource has to be compared with demand. The demand level used here is 1232 LPCD,[2] the average of three different projections in Song *et al.* (2004), shown in Table 2 below. Annex 1 has a brief note examining the projection methodologies.

Table 2:   Water Demand Projection

	2010	2020	
		Scenario 1	Scenario 2
Population (100 million)	13.6	14.4	14.4
GDP (100 billion constant 2000 USD)	28.89	45.88	45.88
Agriculture	3.47	3.67	4.59
Industry	11.79	17.89	18.09
Model 1: Water use by sector			
Agriculture (100 million cu.m)	3,840	3,250	4,000
Industry (100 million cu.m)	1,370	1,570	1,500
Domestic (100 million cu.m)	775	900	
Environment (100 million cu.m)	215	235	
Total (100 million cu.m)	6,200	5,955	6,635
Per Capita (cu.m)	456	414	461
LPCD	**1,249***	**1,133**	**1,262**
Model 2: Per capita water use			
Per Capita (cu.m)	445	430	440
Total (100 million cu.m)	6,060	6,200	6,500
LPCD	**1,219***	**1,178**	**1,205**
Model 3: Regression $y = 2576.1x^{0.1161}$			
($R^2 = 0.9519$) y is water consumption per year in 100 million cubic meter; x is GDP index)			
Total (100 million cu.m)	6,100	6,510	
Per Capita (cu.m)	449	452	
LPCD	**1,229***	**1,239**	

Note: * Average of the three is 1,232.33. The average is used as the norm for IDWA computation in this paper.
Source: Song *et al.*, 2004.

_____

[2] The most commonly accepted view on China's water demand is that it has reached low growth era with the peak of 660,000 million cubic meters coming

Table 3 shows, for each province, the Resource Index, which is the average water resources per capita in LPCD as percentage of the chosen norm for water demand (1,232 LPCD). It is of note that the uniform norm of 1,232 LPCD is used for each province due to lack of provincial demand projections. For the provinces with more than adequate resources, the index is higher than 100, but it is normalized to 100.

Table 3:   Resource Index

Year	Water Resource per capita (cu.m)				Average Water Resource LPCD	Resource Index
	2005	2006	2007	Average		
National	2,151.8	1,932.1	1,916.3	2,000.07	5,480	
Beijing	151.2	141.5	148.2	146.97	403	33
Tianjin	102.2	95.5	103.3	100.33	275	22
Hebei	197	156.1	173.1	175.40	481	39
Shanxi	251.5	263.1	305.6	273.40	749	61
Inner Mongolia	1,917.3	1,719.8	1,232.2	1,623.10	4,447	100
Liaoning	896.3	615.5	610.8	707.53	1,938	100
Jilin	2,066.8	1,300.3	1,269.2	1,545.43	4,234	100
Heilongjiang	1,954.2	1,904.8	1,286.4	1,715.13	4,699	100
Shanghai	138	153.9	187.9	159.93	438	36
Jiangsu	626.6	538.3	653.3	606.07	1,660	100
Zhejiang	2,077.2	1,829.5	1,777.2	1,894.63	5,191	100
Anhui	1,178.8	949.3	1,165.3	1,097.80	3,008	100
Fujian	3,975.5	4,577.7	3,005.7	3,852.97	10,556	100
Jiangxi	3,513.2	3,768.7	2,556.5	3,279.47	8,985	100
Shandong	451	214.8	414.6	360.13	987	80
Henan	597.2	342.8	196.1	378.70	1,038	84

in 2030, when the population maximizes (Wong *et al.*, 2008). Ministry of Water Resources has announced that water demand per capita has reached its plateau and will maintain at around 1,200 LPCD level till 2030. Song *et al.* (2004), from National Development and Reform Commission (NDRC), provide a more evidence based projection through three different approaches, and the results are analogous with that from MWR.

Table 3: Continued

Year	Water Resource per capita (cu.m)				Average Water Resource LPCD	Resource Index
	2005	2006	2007	Average		
Hubei	1,640.6	1122	1,782.1	1,514.90	4,150	100
Hunan	2,649.5	2,794.9	2,247.1	2,563.83	7,024	100
Guangdong	1,906.4	2,396.1	1,686.3	1,996.27	5,469	100
Guangxi	3,703.8	4,011.3	2,922.4	3,545.83	9,715	100
Hainan	3,722.4	2,735.4	3,373.3	3,277.03	8,978	100
Chongqing	1,827.4	1,356.8	2,357.6	1,847.27	5,061	100
Sichuan	3,569.6	2,278.1	2,822.6	2,890.10	7,918	100
Guizhou	2,244.4	2,176.1	2,805.2	2,408.57	6,599	100
Yunnan	4,161.7	3,832.2	5,013.9	4,335.93	11,879	100
Tibet	16,117.06	149,001.4	152,969.2	106,029.22	290,491	100
Shaanxi	1,322.7	739.1	1,007.7	1,023.17	2,803	100
Gansu	1,042.4	709.9	875.9	876.07	2,400	100
Qinghai	16,176.9	10,430.8	12,029.5	12,879.07	35,285	100
Ningxia	143.6	176.8	171.1	163.83	449	36
Xinjiang	4,808.9	4,695.1	4,167.8	4,557.27	12,486	100

Source: *China Statistical Yearbook* 2006, 2007 and 2008.

Access

In the first attempt to compute IDWA (ADB, 2007) access is measured as percentage of population with access to "sustainable improved water sources", which is defined by the WHO and UNICEF's Joint Monitoring Programme as "house connections, public standpipes, boreholes, protected dug wells, protected springs or rainwater collections" (WHO, 2002). In the estimation of IDWA for the states of India (Rao, 2009), water supply via tap, hand-pump and tube-well were all considered as access to improved water sources. In respect of China, data available for the urban areas are exclusively on house connections of tap water. For rural areas, however, data are available both for access to improved water sources and access via house connections. The urban and rural data sources are explained in Box 3.

Box 3

Data Sources for Urban and Rural Access to Water

China Urban Household Sample Survey

China Urban Household Sample Survey is carried out on a quarterly basis to monitor the living conditions of urban population in China. In the Urban Household Basic Conditions Questionnaire, two questions are asked regarding drinking water and water use. One pertains to the household source of drinking water: from tap, mineral water, purified water and well/river; and the other pertains to domestic water source: from home connection, shared tap, well/river and others. The survey uses multi-stage random sampling with a sample size of 36,000 households. A third of the sample is replaced every year. The data from this survey are published annually in the provincial statistical yearbooks.

Source: NBS (2008b), *Regulation on Statistics 2008*.

Second Agriculture Census in China

The Second Agriculture Census of China, source of data on drinking water for rural areas was conducted in 2007 by the National Bureau of Statistics. Questions relating to drinking water were incorporated in the family household questionnaire and comprise drinking water sources in general, drinking water difficulty if any and house connections if available. The 'administrative village questionnaire' collects information on the main source for drinking water at the level of the village. The data are published on household basis in the Synthesis of Second National Agriculture Census.

Source: Second National Agricultural Census Committee of the State Council (2008), *Synthesis of Second National Agriculture Census*.

Based on the data available, Access-I and Access-II are defined as follows:

Access-I:

$100 \times$ (*Urban access rate via house connections* $\times Pu$ *+ Rural access rate to improved water sources* $\times Pr$)

Access-II:

$$100 \times (Urban\ access\ rate\ via\ house\ connections \times Pu$$
$$+ Rural\ access\ rate\ via\ house\ connections \times Pr)$$

where Pu and Pr stand for urban and rural population proportion respectively.

Tables 4 and 5 provide the estimates for Access-I and Access-II respectively.

Table 4:　Access Index-I

Year	Percentage of urban households with house connections	Percentage of urban households	Rural households with access to improved water sources[3]	Percentage of rural households	Percentage of total households with general access	Access Index-I
	2007	2007	2007	2007	2007	
National	97.7	31.84	94.1	68.16	95.2	
Beijing	98.3	75.35	99.8	24.65	98.7	98.7
Tianjin	98.9	67.42	99.4	32.58	99.1	99.1
Hebei	99.0	23.47	98.9	76.53	98.9	98.9
Shanxi	97.6	30.26	96.2	69.74	96.6	96.6
Inner Mongolia	97.0	40.12	99.4	59.88	98.5	98.5
Liaoning	99.6	51.77	99.9	48.23	99.7	99.7
Jilin	97.0	49.97	99.8	50.03	98.4	98.4
Heilongjiang	98.3	55.68	99.9	44.32	99.0	99.0
Shanghai	100.0	81.49	99.9	18.51	100.0	100.0
Jiangsu	99.5	35.95	99.5	64.05	99.5	99.5
Zhejiang	99.9	33.26	93.6	66.74	95.7	95.7
Anhui	98.4	18.65	95.3	81.35	95.9	95.9
Fujian	99.3	29.38	92.6	70.62	94.5	94.5
Jiangxi	97.3	23.21	93.9	76.79	94.7	94.7
Shandong	91.8	22.34	99.4	77.66	97.7	97.7
Henan	95.8	19.56	98.5	80.44	98.0	98.0
Hubei	98.9	33.47	88.2	66.53	91.8	91.8
Hunan	99.3	23.94	94.0	76.06	95.3	95.3

continued overleaf

---

[3] Improved water sources included here are: tap water, deep well, shallow well and rain water collection. Please refer to Annex 2 for the breakdown of data on households' drinking water sources.

Table 4: Continued

Year	Percentage of urban households with house connections	Percentage of urban households	Rural households with access to improved water sources[3]	Percentage of rural households	Percentage of total households with general access	Access Index-I
	2007	2007	2007	2007	2007	
Guangdong	97.2	48.84	91.7	51.16	94.4	94.4
Guangxi	96.2	18.30	81.0	81.70	83.8	83.8
Hainan	95.1	32.21	96.8	67.79	96.2	96.2
Chongqing	95.3	23.15	89.1	76.85	90.6	90.6
Sichuan	96.9	19.39	92.5	80.61	93.3	93.3
Guizhou	96.9	14.84	94.2	85.16	94.6	94.6
Yunnan	96.9	21.84	82.1	78.16	85.3	85.3
Tibet	92.5	19.18	17.7	80.82	32.0	32.0
Shaanxi	96.8	23.52	90.1	76.48	91.7	91.7
Gansu	99.2	25.46	85.8	74.54	89.2	89.2
Qinghai	98.3	39.70	64.3	60.30	77.8	77.8
Ningxia	98.7	36.09	91.8	63.91	94.3	94.3
Xinjiang	91.8	44.19	85.1	55.81	88.0	88.0

Source of basic data: Provincial Statistical Yearbook 2008; and Synthesis of Second National Agriculture Census.

The correlation coefficient between the indicators of Access-I and Access-II is only 0.4. Thus, Access-II is a more suitable indicator since 100% house connections is the target for most of the provinces in both urban and rural areas of China.

Table 5: Access Index-II

Year	Percentage of urban households with house connections	Total urban households	Percentage of rural households with house connections	Total rural households	Percentage of total households with house connections	Access Index-II
	2007	2007	2007	2007	2007	
National	97.7	31.84	48.6	68.16	64.2	
Beijing	98.3	75.35	97.00	24.65	98.0	98.0
Tianjin	98.9	67.42	84.80	32.58	94.3	94.3

Table 4: Continued

Year	Percentage of urban households with house connections 2007	Total urban households 2007	Percentage of rural households with house connections 2007	Total rural households 2007	Percentage of total households with house connections 2007	Access Index-II 2007
Hebei	99.0	23.47	68.40	76.53	75.6	75.6
Shanxi	97.6	30.26	68.10	69.74	77.0	77.0
Inner Mongolia	97.0	40.12	32.60	59.88	58.4	58.4
Liaoning	99.6	51.77	45.50	48.23	73.5	73.5
Jilin	97.0	49.97	31.90	50.03	64.4	64.4
Heilongjiang	98.3	55.68	42.80	44.32	73.7	73.7
Shanghai	100.0	81.49	98.30	18.51	99.7	99.7
Jiangsu	99.5	35.95	83.00	64.05	88.9	88.9
Zhejiang	99.9	33.26	83.40	66.74	88.9	88.9
Anhui	98.4	18.65	19.50	81.35	34.2	34.2
Fujian	99.3	29.38	73.70	70.62	81.2	81.2
Jiangxi	97.3	23.21	17.60	76.79	36.1	36.1
Shandong	91.8	22.34	62.80	77.66	69.3	69.3
Henan	95.8	19.56	27.70	80.44	41.0	41.0
Hubei	98.9	33.47	28.60	66.53	52.1	52.1
Hunan	99.3	23.94	25.60	76.06	43.2	43.2
Guangdong	97.2	48.84	58.70	51.16	77.5	77.5
Guangxi	96.2	18.30	41.20	81.70	51.3	51.3
Hainan	95.1	32.21	37.30	67.79	55.9	55.9
Chongqing	95.3	23.15	33.50	76.85	47.8	47.8
Sichuan	96.9	19.39	35.10	80.61	47.1	47.1
Guizhou	96.9	14.84	52.20	85.16	58.8	58.8
Yunnan	96.9	21.84	57.50	78.16	66.1	66.1
Tibet	92.5	19.18	8.70	80.82	24.8	24.8
Shaanxi	96.8	23.52	49.00	76.48	60.2	60.2
Gansu	99.2	25.46	36.70	74.54	52.6	52.6
Qinghai	98.3	39.70	54.30	60.30	71.7	71.7
Ningxia	98.7	36.09	25.30	63.91	51.8	51.8
Xinjiang	91.8	44.19	63.60	55.81	76.1	76.1

Source: Same as for Table 4.

Capacity

Capacity, on the one hand, is the ability to buy water and on the other, it is the ability of the province to improve water infrastructure. Per capita provincial GDP is used as the indicator of capacity in IDWA. GDP per capita data from 2005 to 2007 are obtained from the annual issues of the *China Statistical Yearbook* (see Box 4 for details on the estimation of GDP).

Box 4

China's Gross Domestic Product Estimation

As Xu (2004) explains, GDP is estimated quarterly and annually, both at provincial and national levels by the Bureau of Statistics. Three main data sources are used in the GDP estimation:

(a) Statistics collected on agriculture, forestry, manufacturing, trade, investment etc., by statistics bureau and other government agencies under the State Council (i.e. on customs, transport, balance of payment etc.); (b) Data from administrative records on final fiscal accounting of revenues and expenditures; and (c) Data from final financial accounts by institutions such as banks, insurance companies, etc.

GDP estimates are released in three stages. For any given year 'n', releases are (a) *preliminary* estimates in February of (n + 1) in *China's Statistical Communiqué*; (b) revised data in *China Statistical Yearbook* in the second half of (n + 1); and c) final estimates in *China Statistical Yearbook* of year n + 2. For example, the final GDP for 2009 will appear in the *China Statistical Yearbook 2011*.

Among all the provinces, Shanghai has the highest per capita GDP and is taken in log form [4.77] as 100, while Guizhou has the lowest and its log [3.77] is taken as 0. The indexes for other provinces are computed as

$$100 \times \frac{log\ (provincial\ per\ capita\ GDP) - 3.77}{4.77 - 3.77}$$

Table 6 presents the results for capacity index.

Quality

Diarrhoeal death rate is used as an indicator of quality in other IDWA papers. Globally, diarrhoea is the leading cause of illness

Table 6: Capacity Index

	GDP per capita (yuan)				Log of Average GDP per capita	Capacity Index
	2005	2006	2007	Average		
National	14,170	16,155	18,665	16,330	4.21	
Beijing	45,444	50,467	58,204	51,372	4.71	94
Tianjin	35,783	41,163	46,122	41,023	4.61	85
Hebei	14,782	16,962	19,877	17,207	4.24	47
Shanxi	12,495	14,123	16,945	14,521	4.16	39
Inner Mongolia	16,331	20,053	25,393	20,592	4.31	54
Liaoning	18,983	21,788	25,729	22,167	4.35	58
Jilin	13,348	15,720	19,383	16,150	4.21	44
Heilongjiang	14,434	16,195	18,478	16,369	4.21	44
Shanghai	51,474	57,695	66,367	58,512	4.77	100
Jiangsu	24,560	28,814	33,928	29,101	4.46	70
Zhejiang	27,703	31,874	37,411	32,329	4.51	74
Anhui	8,675	10,055	12,045	10,258	4.01	24
Fujian	18,646	21,471	25,908	22,008	4.34	57
Jiangxi	9,440	10,798	12,633	10,957	4.04	27
Shandong	20,096	23,794	27,807	23,899	4.38	61
Henan	11,346	13,313	16,012	13,557	4.13	36
Hubei	11,431	13,296	16,206	13,644	4.13	36
Hunan	10,426	11,950	14,492	12,289	4.09	32
Guangdong	24,435	28,332	33,151	28,639	4.46	69
Guangxi	8,788	10,296	12,555	10,546	4.02	25
Hainan	10,871	12,654	14,555	12,693	4.10	33
Chongqing	10,982	12,457	14,660	12,700	4.10	33
Sichuan	9,060	10,546	12,893	10,833	4.03	26
Guizhou	5,052	5,787	6,915	5,918	3.77	0
Yunnan	7,835	8,970	10,540	9,115	3.96	19
Tibet	9,114	10,430	12,109	10,551	4.02	25
Shaanxi	9,899	12,138	14,607	12,215	4.09	32
Gansu	7,477	8,757	10,346	8,860	3.95	18
Qinghai	10,045	11,762	14,257	12,021	4.08	31
Ningxia	10,239	11,847	14,649	12,245	4.09	32
Xinjiang	13,108	15,000	16,999	15,036	4.18	41

Source: *China Statistical Yearbook* 2006, 2007 and 2008.

and death, and 88% of diarrhoeal deaths can be attributed to unsafe water supply, sanitation and hygiene, and is mostly concentrated in children in developing countries (WHO, 2004). However, in China, the diarrhoeal death rate was as low as 0.085 and 0.1 cases per million people in 2006 and 2007 respectively (MOH, 2008). This speaks well about the control and almost the ending of death due to diarrhoea. We, therefore, need to choose an alternative quality indicator.

The World Bank (2006) studied environmental cost of pollution in China, in which it calculated the amount of surface water supplied for domestic use that does not meet drinking water standards.[4] This is used as a proxy for quality in this paper. The quality index is computed as follows and the results are in Table 7.

$$100 - \frac{Polluted\ water\ supplied\ for\ domestic\ use\ (in\ LPCD)}{Total\ water\ supplied\ for\ domestic\ use\ (in\ LPCD)}$$

Table 7: Quality Index

	Polluted water supplied for domestic use (million cubic meter)	Total water supplied for domestic use (million cubic meter)	Population (year end, 10,000 persons)	Polluted water supplied for domestic use (LPCD)	Total water supplied for domestic use (LPCD)	Polluted water as percentage of total domestic water supplied	Quality Index
Year	2006	2006	2006	2006	2006	2006	
Beijing	0.00	14.4	1,581	0.00	250	0.00	100.0
Tianjin	0.00	4.6	1,075	0.00	117	0.00	100.0
Hebei	0.00	24.1	6,898	0.00	96	0.00	100.0
Shanxi	99.00	9.3	3,775	0.77	67	1.15	99.9
Inner Mongolia	13.00	13.1	2,397	0.15	150	0.10	100.0
Liaoning	41.00	24.3	4,271	0.26	156	0.17	100.0
Jilin	309.00	11.5	2,723	3.11	116	2.68	99.7

---

[4] See Annex 2 for drinking water standards in China. The World Bank study considers water which does not meet Class III standards as not meeting the drinking water standards.

Table 7:   Continued

	Polluted water supplied for domestic use (million cubic meter)	Total water supplied for domestic use (million cubic meter)	Population (year end, 10,000 persons)	Polluted water supplied for domestic use (LPCD)	Total water supplied for domestic use (LPCD)	Polluted water as percentage of total domestic water supplied	Quality Index
Year	2006	2006	2006	2006	2006	2006	
Heilongjiang	289.00	20.0	3,823	2.07	143	1.45	99.9
Shanghai	1,258.00	20.4	1,815	18.97	308	6.16	99.4
Jiangsu	1,316.00	46.1	7,550	4.78	167	2.86	99.7
Zhejiang	594.00	32.6	4,980	3.27	179	1.83	99.8
Anhui	387.00	24.4	6,110	1.73	109	1.59	99.8
Fujian	13.00	21.0	3,558	0.10	162	0.06	100.0
Jiangxi	4.00	20.9	4,339	0.03	132	0.02	100.0
Shandong	54.00	31.3	9,309	0.15	92	0.16	100.0
Henan	231.00	34.6	9,392	0.67	101	0.66	99.9
Hubei	106.00	28.8	5,693	0.51	139	0.37	100.0
Hunan	575.00	44.2	6,342	3.78	191	1.98	99.9
Guangdong	1,188.00	92.4	9,304	3.49	272	1.28	99.9
Guangxi	72.00	41.9	4,719	0.42	243	0.17	100.0
Hainan	0.00	5.9	836	0.00	193	0.00	100.0
Chongqing	138.00	16.2	2,808	1.35	158	0.85	99.9
Sichuan	373.00	34.2	8,169	1.25	115	1.09	99.9
Guizhou	1.00	17.7	3,757	0.01	129	0.01	100.0
Yunnan	0.00	19.5	4,483	0.00	119	0.00	100.0
Tibet	0.00	2.5	281	0.00	244	0.00	100.0
Shaanxi	7.00	13.3	3,735	0.05	98	0.05	100.0
Gansu	298.00	9.1	2,606	3.13	96	3.26	99.7
Qinghai	13.00	3.2	548	0.65	160	0.41	100.0
Ningxia	2.00	1.8	604	0.09	82	0.11	100.0
Xinjiang	0.00	10.7	2,050	0.00	143	0.00	100.0

Source: *China Statistical Yearbook* 2007; and World Bank (2006).

Use

Per capita water use for domestic purpose is taken as use indicator. Total amount of water used for domestic purpose in each province in 2006 and 2007 is available in *China Statistical Yearbook* for 2007 and 2008 respectively. The data are then converted to LPCD using the population for the two years.

Box 5

Water Use Data

The water use data for domestic sector contains two parts: (a) metered water data: data on domestic water use through metered water supply for both rural and urban areas are collected from utility boards, (b) other sources: data on rural domestic water use via other sources are collected through sample survey. The results are combined and published annually in *China Statistical Yearbook*.

Source: Interview with officials from Ministry of Water Resources.

The Ministry of Housing and Urban-Rural Development (MOHURD, 2002) regulated the quantity of water to be used in urban areas for domestic purposes. The standards (see Annex 4) for different regions are set based on the level of development and resource availability. However, these standards are not used to compute the provincial use index in the belief that every citizen should be entitled with the same optimal amount of water for use. Thus, per capita domestic water use in Singapore in 2007 — 157 LPCD[5] was taken as the norm and pegged at 100 for use index. There is strong reason for using the norm based on Singapore experience. It illustrates how the City State with limited water resources, has ensured that its residents enjoy continuous supply of water that can be consumed straight from the tap.

In the computation of the provincial use index, provinces with actual use higher than 157 LPCD are simply credited with the index of 100. In the standards set by the Chinese government, 75 LPCD is the least requirement for any province to provide for domestic use to its citizens and hence considered as minimum need, which is considered as 0.

---

[5] This is from the Singapore Ministry of Environment and Water Resources (MEWR) (2008).

$$100 - \frac{provincial\ per\ capita\ water\ use - 75}{157 - 75}$$

The result of use index is shown in Table 8.

Table 8: Use Index

	Water for Domestic Use (100 million cu.m)			Population (year-end, 10,000 persons)			Average Water for domestic use LPCD	Use Index
	2006	2007	Average	2006	2007	Average		
National	693.8	710.4	702.1	1,581	1,633	1,607	149	
Beijing	14.4	14.6	14.5	1,075	1,115	1,095	251	100.0
Tianjin	4.6	4.8	4.7	6,898	6,943	6,921	119	54.3
Hebei	24.1	23.9	24.0	3,775	3,393	3,584	95	24.8
Shanxi*	9.3	9.5	9.4	2,397	2,405	2,401	73	0.0*
Inner Mongolia	13.1	14.2	13.7	4,271	4,298	4,285	156	98.9
Liaoning	24.3	24.3	24.3	2,723	2,730	2,727	154	96.0
Jilin	11.5	11.7	11.6	3,823	3,824	3,824	117	50.9
Heilongjiang	20.0	18.6	19.3	1,815	1,858	1,837	138	77.2
Shanghai	20.4	21.6	21.0	7,550	7,625	7,588	317	100.0
Jiangsu	46.1	48.4	47.3	4,980	5,060	5,020	171	100.0
Zhejiang	32.6	33.9	33.3	6,110	6,118	6,114	183	100.0
Anhui	24.4	26.1	25.3	3,558	3,581	3,570	113	46.5
Fujian	21.0	21.2	21.1	4,339	4,368	4,354	162	100.0
Jiangxi	20.9	22.9	21.9	9,309	9,367	9,338	138	77.2
Shandong	31.3	32.5	31.9	9,392	9,360	9,376	94	23.0
Henan	34.6	32.7	33.7	5,693	5,699	5,696	98	28.4
Hubei	28.8	29.4	29.1	6,342	6,355	6,349	140	79.1
Hunan	44.2	44.6	44.4	9,304	9,449	9,377	192	100.0
Guangdong	92.4	90.5	91.5	4,719	4,768	4,744	269	100.0
Guangxi	41.9	48.6	45.3	836	845	841	263	100.0
Hainan	5.9	6.1	6.0	2,808	2,816	2,812	197	100.0
Chongqing	16.2	17.3	16.8	8,169	8,127	8,148	163	100.0
Sichuan	34.2	34.4	34.3	3,757	3,762	3,760	115	48.8
Guizhou	17.7	16.9	17.3	4,483	4,514	4,499	126	62.7
Yunnan	19.5	19.9	19.7	281	284	283	120	55.4
Tibet	2.5	2.1	2.3	3,735	3,748	3,742	225	100.0
Shaanxi	13.3	13.6	13.5	2,606	2,617	2,612	99	28.9
Gansu	9.1	9.5	9.3	548	552	550	98	27.8
Qinghai	3.2	3.3	3.3	604	610	607	163	100.0
Ningxia	1.8	1.8	1.8	2,050	2,095	2,073	82	8.2
Xinjiang	10.7	11.3	11.0	2,006	2,007	129,725	147	87.7

Note: *Actually, the index was a little over negative 2.
Source: *China Statistical Yearbook* 2007 and 2008.

It is important to keep in mind that although some provinces achieved 100 in this index, it might not be appropriate to consider their domestic water use to be as good as Singapore. In Singapore, due to massive water saving measurements, the per capita domestic water use has been constantly declining since the 1990s.[6] But in China, per capita domestic water use is still increasing, which means that the provinces with the relatively very high index are using more water than needed; and there could be a case for conservation measures in the medium term in those regions.

IDWA-I and IDWA-II

IDWA-I is constructed by averaging the resource, access-I, capacity, quality and use indicators. IDWA-II is compiled similarly, but using access-II instead of access-I.

Table 9:   IDWA-I and IDWA-II

	Resource	Access I	Access II	Capacity	Quality	Use	IDWA I	IDWA II
Beijing	33	98.7	98.0	94	100.0	100.0	85.1	85.0
Tianjin	22	99.1	94.3	85	100.0	54.3	72.1	71.1
Hebei	39	98.9	75.6	47	100.0	24.8	61.9	57.3
Shanxi	61	96.6	77.0	39	99.9	0.0	59.3	55.4
Inner Mongolia	100	98.5	58.4	54	100.0	98.9	90.3	82.3
Liaoning	100	99.7	73.5	58	100.0	96.0	90.7	85.5
Jilin	100	98.4	64.4	44	99.7	50.9	78.6	71.8
Heilongjiang	100	99.0	73.7	44	99.9	77.2	84.0	79.0
Shanghai	36	100.0	99.7	100	99.4	100.0	87.1	87.0
Jiangsu	100	99.5	88.9	70	99.7	100.0	93.8	91.7
Zhejiang	100	95.7	88.9	74	99.8	100.0	93.9	92.5
Anhui	100	95.9	34.2	24	99.8	46.5	73.2	60.9
Fujian	100	94.5	81.2	57	100.0	100.0	90.3	87.6
Jiangxi	100	94.7	36.1	27	100.0	77.2	79.8	68.1
Shandong	80	97.7	69.3	61	100.0	23.0	72.3	66.7

---

[6] See Singapore MEWR (2008).

Table 9: Continued

	Resource	Access I	Access II	Capacity	Quality	Use	IDWA I	IDWA II
Henan	84	98.0	41.0	36	99.9	28.4	69.3	57.9
Hubei	100	91.8	52.1	36	100.0	79.1	81.4	73.4
Hunan	100	95.3	43.2	32	99.9	100.0	85.4	75.0
Guangdong	100	94.4	77.5	69	99.9	100.0	92.7	89.3
Guangxi	100	83.8	51.3	25	100.0	100.0	81.8	75.3
Hainan	100	96.2	55.9	33	100.0	100.0	85.8	77.8
Chongqing	100	90.6	47.8	33	99.9	100.0	84.7	76.1
Sichuan	100	93.3	47.1	26	99.9	48.8	73.6	64.4
Guizhou	100	94.6	58.8	0	100.0	62.7	71.5	64.3
Yunnan	100	85.3	66.1	19	100.0	55.4	71.9	68.1
Tibet	100	32.0	24.8	25	100.0	100.0	71.4	70.0
Shaanxi	100	91.7	60.2	32	100.0	28.9	70.5	64.2
Gansu	100	89.2	52.6	18	99.7	27.8	66.9	59.6
Qinghai	100	77.8	71.7	31	100.0	100.0	81.8	80.5
Ningxia	36	94.3	51.8	32	100.0	8.2	54.1	45.6
Xinjiang	100	88.0	76.1	41	100.0	87.7	83.3	81.0

The correlation coefficient between IDWA-I and IDWA-II is 0.96, which is due to the domination of the four common components. IDWA-II, however, should be preferred since house connection is the optimum choice. The map below exhibits the IDWA variations across China. Low drinking water adequacy still prevails in vast tracts of the county.

4. Selected Inter-relationships between IDWA and Other Development Indicators

Table 10 is a compilation of IDWA and other important development indicators from the *China Human Development Report 2007/2008* (UNDP, 2009). The inter-correlations between these indicators and IDWA are noted below for Chinese provinces and the Indian states respectively computed by the author of this paper

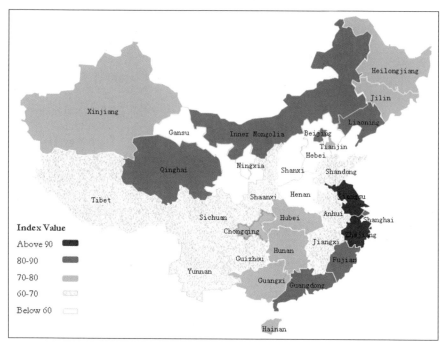

Figure 1: Drinking Water Adequacy across China

and Rao (2009). Despite the considerable superiority of China relative to India on the 5 development indicators, it is significant that
the broad patterns of correlations hold equally.

	China: Correlation with		India: Correlation with	
Development Indicator	*IDWA I*	*IDWA II*	*IDWA I*	*IDWA II*
Income Index	0.46	0.59	0.51	0.62
Human Development Index	0.42	0.48	0.22	0.42
Life Expectancy	0.35	0.37	0.05	0.22
Education Index	0.31	0.31	0.05	0.31
Poverty	−0.42	−0.35	−0.39	−0.50

Reinforcing the findings in IDWA for states of India, the
IDWA-II has a stronger positive correlation with life expectancy,

income and human development. Income index, in both studies has the strongest positive correlation with IDWA as compared to other development indicators. Contrary to the result in IDWA for states of India, IDWA-I in this study has a higher (negative) correlation with the poverty ratio as compared to IDWA-II, a matter that needs further probing.

Table 10: Selected Development Indicators

	IDWA I	IDWA II	Life expectancy index	Education Index	Income Index	Human Development Index	Poverty Ratio
Beijing	85.1	85.0	0.85	0.92	0.92	0.90	0.6
Tianjin	72.1	71.1	0.83	0.93	0.88	0.88	0.1
Hebei	61.9	57.3	0.79	0.87	0.73	0.80	2.6
Shanxi	59.3	55.4	0.78	0.87	0.70	0.78	8
Inner Mongolia	90.3	82.3	0.75	0.83	0.76	0.78	5.6
Liaoning	90.7	85.5	0.81	0.88	0.78	0.82	4.2
Jilin	78.6	71.8	0.80	0.86	0.72	0.80	4.8
Heilongjiang	84.0	79.0	0.80	0.87	0.73	0.80	4.3
Shanghai	87.1	87.0	0.89	0.93	0.94	0.92	0
Jiangsu	93.8	91.7	0.82	0.86	0.84	0.84	0.7
Zhejiang	93.9	92.5	0.82	0.85	0.82	0.83	0.4
Anhui	73.2	60.9	0.78	0.78	0.65	0.74	2.2
Fujian	90.3	87.6	0.79	0.82	0.77	0.80	0.3
Jiangxi	79.8	68.1	0.73	0.84	0.66	0.74	3.7
Shandong	72.3	66.7	0.82	0.84	0.79	0.82	1.1
Henan	69.3	57.9	0.78	0.83	0.69	0.77	2.9
Hubei	81.4	73.4	0.77	0.84	0.69	0.77	2.8
Hunan	85.4	75.0	0.76	0.85	0.68	0.76	3
Guangdong	92.7	89.3	0.81	0.86	0.82	0.83	1.1
Guangxi	81.8	75.3	0.77	0.84	0.65	0.76	3.6
Hainan	85.8	77.8	0.80	0.82	0.68	0.77	0.6
Chongqing	84.7	76.1	0.78	0.83	0.68	0.76	5.3
Sichuan	73.6	64.4	0.77	0.80	0.65	0.74	3.4
Guizhou	71.5	64.3	0.68	0.74	0.55	0.66	9
Yunnan	71.9	68.1	0.68	0.76	0.63	0.69	8.1
Tibet	71.4	70.0	0.66	0.55	0.65	0.62	20.6
Shaanxi	70.5	64.2	0.75	0.84	0.68	0.76	6.7
Gansu	66.9	59.6	0.71	0.73	0.62	0.69	7
Qinghai	81.8	80.5	0.68	0.75	0.67	0.70	12.6
Ningxia	54.1	45.6	0.75	0.79	0.67	0.74	9.6
Xinjiang	83.3	81.0	0.71	0.84	0.71	0.75	7.7

Source: UNDP (2009), *China Human Development Report 2007/2008.*

5. Concluding Remarks

Institutional Implications

At the national level, various ministries are involved in different aspects governing drinking water supply. The National Development and Reform Commission (NDRC) and MWR are the leading agencies in overall water sector planning while MWR, MOHURD, Ministry of Health (MOH) and Ministry of Environmental Protection (MEP) each carrying out certain functions related to drinking water supply, as demonstrated in Figure 2.

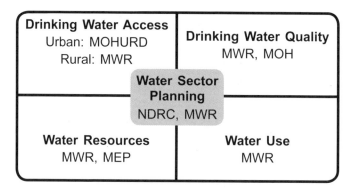

Figure 2: Governance Structure of Drinking Water Supply

It is easy to see that the different components of IDWA are under various ministries' spheres and some components fall into several ministries' purview. Therefore, the improvement of overall drinking water adequacy requires strengthened inter-ministerial (central level) and inter-departmental (provincial level) cooperation.

Provincial Variation

IDWA reconfirms that the reasons for low drinking water adequacy vary from province to province. Figure 3 takes Shanxi, Anhui and Henan as examples to demonstrate the differences. Shanxi is suffering from low per capita water for domestic use and lack of resources; Henan's access rate and per capita water for domestic use are both low; while Anhui is constrained by low capacity and low water access, despite its relatively high water use and enough resources.

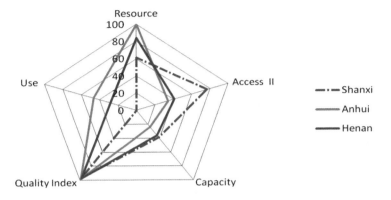

Figure 3: Components of IDWA

The case is strong for regionally focused measures to address the typical constraints that are regional in nature. Addressing water resource shortage, for example, might require a provincial government to choose to increase supply by investing in *water transfer projects*, reduce the demand through water saving measures or implement both, since water transfer projects could be costly and may not be viable due to resource unavailability. In recent years, Beijing is getting water from its neighbouring Hebei province, which is also suffering from water scarcity. Though the project is praised for releasing the water pressure in Beijing, it might not be a sustainable measure considering the potential damage it can cause to the people, economy and environment in Hebei.

The longer term solutions have to be based on serious discourse on inter-provincial differences and coordinated policies and programs. IDWA or its variants based on further modifications and simplifications could serve to alert the central and provincial governments of China on the gaps to be filled and differences to be bridged by coordinated efforts for all to be well served with appropriate quantity and quality of water.

Data Accuracy

Data accuracy issue is critical for proper index computing, which alone assists in monitoring differences and progress in reducing them. Problems such as ambiguous definition of indicators, inconsistent calibration used by different organizations or by same organization in different years could make comparisons less useful. For example, water coverage rate is explained as population with access to water supply facilities in the *China Statistical Yearbook*. By water supply facilities, it refers to only public water plants and by access, it means access through metered tap. Another example is per capita based data since the definition of population employed might be very different: population registered with police office versus population actually residing, year-end population versus annual average population etc. which are sometimes used interchangeably without clear indication to that effect. Standardized concepts and definitions should be adopted by all agencies across the board.

Fine-tuning data collection and reporting systems of water related data (or any data for that matter) are continuous processes and must be treated as such by the agencies concerned. Then only it becomes possible to ensure proper monitoring.

Revisiting IDWA

China's water sector continues to transform rapidly as the economy progresses. The first phase of the Central Route of South-to-North Water Diversion Project will start to supply water to Beijing in 2014, which will dramatically alter the water resource availability across China. More and more water saving initiatives, water reuse and waste water treatment technologies are also recreating the dynamics of water sector. This implies the need for revisiting IDWA in order to provide timely insights for policy adjustment. It is hoped that this initiative will graduate from this paper to routine government monitoring, perhaps by the NDRC, the designated author of the Plans of Water Development.

References

Asian Development Bank (2007). *Asian Water and Development Outlook.* Manila: Asian Development Bank.

General Administration of Quality Supervision, Inspection and Quarantine of People's Republic of China and Standardization Administration of People's Republic of China (2009). *Code of Practice for Water Resource Bulletin.* Beijing: Standards Press of China.

Hainan Bureau of Water Resources (2006). *Definition of Drinking Water Safety and Drinking Water Difficulty.* Available at <http://swj.hainan.gov.cn/rxys1. htm> [Accessed January 22, 2010].

National Bureau of Statistics (2006). *China Statistical Yearbook 2006.* Beijing: China Statistics Press.

———— (2007). *China Statistical Yearbook 2007.* Beijing: China Statistics Press.

———— (2008a). *China Statistical Yearbook 2008.* Beijing: China Statistics Press.

———— (2008b). *Regulation on Statistics 2008.* Beijing: China Statistics Press.

———— (2009). *China Statistical Yearbook 2009.* Beijing: China Statistics Press.

National Development and Reform Commission, Ministry of Water Resources and Ministry of Construction (2007). *11th Five Year Plan on Water Development.* Available at <http://www.sdpc.gov.cn/ncjj/zhdt/W020070608354115302662.pdf> [Accessed January 12, 2010].

Ministry of Environment and Water Resources of Singapore (2008). *Conserve, Value and Enjoy.* Available at <http://app.mewr.gov.sg/web/Contents/Contents.aspx?ContId=961> [Accessed November 11, 2009].

Ministry of Health (2008). *China Health Statistical Yearbook 2007.* Beijing: Peking Union Medical College Press.

Ministry of Housing and Urban-Rural Construction (2002). *The standard of water quantity for city's residential use.* Available at <http://www.chinajsxy.com/hygfupload/up2007928163436.pdf> [Accessed November 11, 2009].

Ministry of Water Resources (2009). *Minister's Speech on the Celebration of 60th Anniversary of People's Republic of China.* Available at <http://www.mwr.gov.cn/zwzc/ldxx/cl/zyjh/200909/t20090930_123159.html> [Accessed October 22, 2009].

Seetharam, K.E. and B. Rao (2009). Index of Drinking Water Adequacy (IDWA) for the States of India, *Journal of Infrastructure Development*, 1, 179–92.

———— (2010). Index of Drinking Water Adequacy for the Asian Economies, *Water Policy*, 12, Supplement 1, 135–54.

Song, J., Zhang, Q. and Liu, Y. (2004). Analysis of Safety Factors of Water Resources in China in 2020 and Strategic Suggestions, *China Water Resources*, 2004 (9), 14–7.

State Council Second National Agriculture Census Committee and National Bureau of Statistic (2008). *Synthesis of Second National Agriculture Census*. Beijing: China Statistics Press.

United Nation's Development Programme (2009). *China Human Development Report 2007/2008*. Available at <http://hdr.undp.org/en/reports/nationalreports/asiathepacific/china/China_2008_en.pdf> [Accessed December 9, 2009].

Wang, H., Qin, D. and Wang, D. (2008). *Harmonious Development of Water Sector and Economy*, Beijing: China Water Power Press.

World Bank (2007). Water Scarcity and Pollution. In *Cost of Pollution in China: Economic Estimates of Physical Damages*, Bangkok: East Asia and Pacific Region, World Bank.

World Health Organization (2002). *Health through safe drinking water and basic sanitation*. Available at <http://www.who.int/water_sanitation_health/mdg1/en/index.html > [Accessed December 9, 2009].

World Health Organization (2004). *Water, Sanitation and Hygiene links to Health*. Available at <http://www.who.int/water_sanitation_health/publications/facts2004/en/index.html> [Accessed November 26, 2009].

Xinhua News Agency (2008). *Review: 30 Years of Rural Drinking Water Project*. Available at <http://news.xinhuanet.com/newscenter/2008–10/11/content_10179667.htm> [Accessed on January 22, 2010].

Xu, X. (2004). China's Gross Domestic Product Estimation, *China Economic Review*, 15 (3): 302–22.

Annexes

Annex 1: Water Demand in China

Over the past decade, the Chinese government realized the importance of water saving and took out appropriate measures to reduce water use per unit of GDP. These measures included introducing new irrigation methods and increasing water price for both industry and domestic sector to discourage excessive use. The results have been noticeable. Between 1997 and 2002, water use fell but has gone up since, though not to the level of 1997. Overall, per capita water use was more or less fluctuating around 1200 LPCD, in contrast to the continuously fast-paced economic growth, as shown in the figure below. MWR has further estimated that till 2030, the per capita water demand will remain at the same level, at around 1200 LPCD.

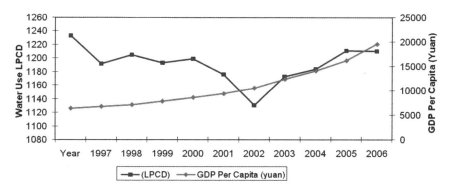

Water use growth rate and economic growth rate

Demand Projections

Song *et al.* (2004) from the National Development and Reform Commission (NDRC) used three different approaches to project water demand in China. Their projection, especially its Model of Water Use by Sector is examined here by comparing water efficiency in large agricultural countries in the world, namely China, USA, Australia, India and Brazil. The water efficiency in terms of water use per unit value added in agriculture and industry sector is calculated for each country for the year 2000 (limited by data availability). The water efficiency for China in 2010 and 2020 in the projected scenario is also calculated. The results are shown in the following table.

Water Efficiency in Agriculture and Industry in Selected Countries

Year	Country	Water Use ($10^9 m^3$ and %) Agriculture		Industry		GDP (10^6 constant 2000 USD) Agriculture	Industry	water efficiency per USD one thousand GDP Cu M Agriculture	Industry
2000	Brazil	36.63	61.77%	10.65	17.96%	31.29	154.86	1.171	0.069
2000	India	558.39	86.46%	35.20	5.45%	98.41	110.35	5.674	0.319
2000	USA	197.75	41.26%	220.71	46.05%	112.1	2,198.1	1.764	0.100
2000	Australia	18.01	75.26%	2.40	10.03%	13.07	100.15	1.378	0.024
2000	China	426.85	67.72%	161.98	25.70%	188.53	550.30	2.364	0.294
2010	China	384.00	61.94%	137.00	22.10%	347	1,179	1.107	0.116
2020	China	325.14	54.60%	157.21	26.40%	367	1,789	0.886	0.088
		400.09	60.30%	149.95	22.60%	459	1,809	0.872	0.083

The table shows that China's water efficiency projected in 2010 in reaching the 2000 level of the other countries (i.e. USA and Brazil), which suggests high possibility of the projection coming true. The 11th Five Years Plan of China has set the target of 30% reduction in water intensity in industry sector. By the end of 2007, 16% of reduction has already been achieved, 4% more than planned, which again implies that the projections noted in the table can indeed be realized.

Annex 2: Drinking Water Standards in China

Drinking Water Pollutants	Class I	Class II	Class III
Chrome (degree)	15.0	20.0	30.0
Turbidity (degree)	3.0	10.	020.0
Total dissolved solids (mg/L, CaCO$_3$)	450.0	550.0	700.0
Iron (mg/L)	0.3	0.5	1.0
Manganese (mg/L)	0.1	0.3	0.5
COD (mg/L)	3.0	6.0	6.0
Chlorate (mg/L)	250.0	300.0	450.0
Sulfate (mg/L)	250.0	300.0	400.0
Fluoride (mg/L)	1.0	1.2	1.5
Arsenic (mg/L)	0.1	0.1	0.1
Nitrate (mg/L)	20.0	20.0	20.0
Total bacteria (/mL)	100.0	200.0	500.0
Total coliform (/L)	3.0	11.0	27.0

Source: World Bank, 2007.

Annex 3: Access to Water Sources for Rural China

					Number of Households
Province	Tap water	deep well	shallow well	rain water collection	Improved water sources sources
National	51,006,384	92,313,119	61,503,510	3,159,848	207,982,861
Beijing	901,466	498,742	33,302	123	1,433,633
Tianjin	287,547	831,104	33,648	69	1,152,368
Hebei	1,875,009	10,198,637	1,476,242	51,799	13,601,687
Shanxi	1,028,169	3,114,134	1,237,455	283,325	5,663,083
Inner Mongolia	433,636	2,308,443	664,476	22,948	3,429,503
Liaoning	1,106,746	4,557,394	770,838	1,827	6,436,805
Jilin	312,982	2,928,570	346,712	1	3,588,265
Heilongjiang	707,715	3,317,314	158,106	259	4,183,394
Shanghai	1,184,305	7,921	18,383	4	1,210,613
Jiangsu	10,798,679	2,483,081	828,148	463	14,110,371
Zhejiang	5,272,941	951,797	2,066,631	35,074	8,326,443
Anhui	1,352,031	7,780,148	2,199,637	10,849	11,342,665
Fujian	1,934,980	1,118,108	2,090,321	7,314	5,150,723

continued overleaf

Annex 3:   Continued

				Number of Households	
Province	*Tap water*	*deep well*	*shallow well*	*rain water collection*	*Improved water sources sources*
Jiangxi	531,757	2,345,418	3,597,840	17,990	6,493,005
Shandong	7,615,314	10,987,820	2,265,578	9,474	20,878,186
Henan	710,227	15,134,603	2,782,827	122,957	18,750,614
Hubei	1,288,082	2,447,052	4,098,483	206,411	8,040,028
Hunan	768,444	3,719,966	6,579,160	72,531	11,140,101
Guangdong	4,739,167	2,059,347	2,802,025	8,131	9,608,670
Guangxi	1,034,497	2,138,740	3,257,389	258,896	6,689,522
Hainan	139,781	389,081	391,710	315	920,887
Chongqing	714,428	837,273	3,323,997	170,879	5,046,577
Sichuan	917,333	6,204,868	7,039,892	195,292	14,357,385
Guizhou	683,392	218,221	5,386,816	236,489	6,524,918
Yunnan	1,491,917	609,488	4,342,245	305,717	6,749,367
Tibet	14,863	23,092	31,556	2,652	72,163
Shaanxi	840,149	2,720,220	2,046,145	335,254	5,941,768
Gansu	780,406	1,278,234	1,075,565	663,930	3,798,135
Qinghai	161,763	97,406	223,625	15,498	498,292
Ningxia	188,390	393,416	110,505	112,899	805,210
Xinjiang	1,190,268	613,481	224,253	10,478	2,038,480

Source: Synthesis of Second National Agriculture Census.

Annex 4:   The Standard of Water Quantity for City's Residential Use

	Quantity (LPCD)	*Provinces*
1	80–135	Heilongjiang, Jilin, Liaoning, Inner Mongolia
2	85–140	Beijing, Tianjin, Hebei, Shandong, Henan, Shanxi, Shaanxi, Ningxia, Gansu
3	120–180	Shanghai, Jiangsu, Zhejiang, Fujian, Jiangxi, Hubei, Hunan, Anhui
4	150–220	Guangdong, Guangxi, Hainan
5	100–140	Chongqing, Sichuan, Guizhou, Yunnan
6	75–125	Xinjiang, Tibet, Qinghai

Source: MOHURD, 2002.

6

Data on Drinking Water Access: The Case of Vietnam

Ngo Quang Vinh

1. Introduction

The Index of Drinking Water Adequacy (IDWA) is based on averaging five component indexes of which one key component refers to "access". Access data are used widely in the Millennium Development Goals (MDGs) and are based on the estimates reported in the database from the Joint Monitoring Program for Water Supply and Sanitation (JMP). JMP was formulated by the World Health Organization (WHO) and the United Nations Children's Fund (UNICEF) to monitor the progress towards the targets relating to drinking water and sanitation. The JMP's mission is "to be the trusted source of global, regional and national data on sustainable access to safe drinking water and basic sanitation, for use by governments, donors, international organizations and civil society".

Since "access" data are critical to proper monitoring of MDGs, it is only fair and appropriate that an evaluation of the JMP database is attempted.

133

Taking Vietnam as a case study and data on drinking water access from the JMP as an example, this paper will briefly outline how JMP collects the data on drinking water access and then summarize a few critical points on advantages and disadvantages of the JMP methodology in general. Finally, an insight into Vietnam data are provided on the basis of analyzing the pertinent data from the JMP database as well as comparing the data on Vietnam with those of China and Thailand, which is to serve to demonstrate how much more can in fact be learnt from a wide-ranging cross-country study of JMP data and methods.

2. The Joint Monitoring Program (JMP) Data Sources and Definitions

Currently JMP has a huge database of nearly 900 nationally representative household surveys and censuses from about 190 countries. JMP data sources include national censuses and surveys, Demographic and Health Surveys (DHS), Multiple Indicator Cluster Surveys (MICS), World Health Surveys (WHS) and Living Standard Measurement Surveys (LSMS).[1]

As definitions of "access" can vary widely within and among countries and regions, and as JMP is mandated to report at global level and across time, it must use a set of categories for "improved drinking water sources" from WHO to analyze the national data on which its trends and estimates are based. "Improved drinking water sources" under WHO category include piped water into dwelling (or house connection), piped water to yard/plot, public tap or standpipe, tube-well/borehole, protected dug-well/spring, bottled water (in some cases) and rain water collection. "Access to water" is broadly defined as the availability of at least 20 litres per person per day from a source within one kilometer of the user's dwelling. The instructions in more recent national surveys in some countries even clearly stated that piped systems should not be considered

---

[1] See Annex for a brief write-up on DHS, MICS, WHS and LSMS.

"functioning" unless they were operating at over 50% capacity on a daily basis and that hand-pumps should not be considered "functioning" unless they were operating for at least 70% of the time with a maximum of two weeks' delay between breakdown and repair. Furthermore, in line with the MDG indicator definition that stipulates "use of improved water sources" to measure "access to improved water sources", JMP measures and reports on the actual use of improved drinking water. It is also worth noting that the household surveys and censuses on which JMP relies also measure "use" and not "access" — since access involves many additional criteria other than use.

3. JMP Methodology

For each country, data from surveys and censuses are plotted on a timescale from 1980 to the present. With these data points, a linear trend line based on least squares is drawn and from this line estimates are made for 1990, 1995, 2000, 2005 and 2008 (if possible). The total estimates are population weighted average of the urban and rural numbers. When necessary, JMP extrapolates the linear regression line up to two years before or after the earliest or latest data point. Outside these time limits, the extrapolated regression line is flat for up to four years. If the extrapolated regression line reaches 0% or 100% coverage or beyond, a flat line is drawn from the year prior to the year where estimates would reach 100% (or 0%). Where insufficient data exists for linear regression, the slope of the regression is assumed to be zero, which means no progress is made. In this case, the projection can be made up to a maximum of six years forward or backwards in time from the data point. When the use of improved drinking water is at 95% or above, or at 5% or below, the projection can be made without limitations.

Advantages of the JMP methodology

With a huge database from most countries in the world, the first and foremost advantage of JMP methodology is that it can allow for inter-country and intra-country comparison. We know that countries use different survey questionnaires or indicators for data collection, hence the collected data at country level do not tell us equivalent story in each country or between countries. Inter-country and intra-country comparability may help us judge or sense what dataset is more reliable and should be used in the estimates. The next advantage of the huge and frequently updated database is that accuracy of the JMP estimates will definitely improve overtime.

The dominant advantage of the JMP methodology is that it can create a trend line for estimation and tell us the approximate data for the years when no survey/census is taken. This is really crucial in monitoring the progress toward MDG targets as the trend line can visualize the path we have gone through and the destination we may reach by the year 2015; and policy makers will know how to proceed if they are determined to meet the MDG targets.

Disadvantages of the JMP methodology

Though the database used by JMP is huge and frequently updated, given that the sources over time could be widely varying, indicators are not always identical and comparable. In addition, JMP just uses the data as they are, without adjusting them to conform to any common benchmarks and yardsticks. This fact may lead to misleading inferences from cross-national comparisons.

Though JMP keeps data from 1980, it uses the data for 1990 as the base-line to draw a trend-line and extrapolate to 2015. In many countries, however, data prior to 1997 are not available and there-fore the trend-line is made from only two or three observations in the most recent decade. The information from this kind of trend-line will not be sufficient to help monitor the progress toward the MDG targets by 2015 on an internationally comparable and nationally consistent basis.

4. Case Study of Vietnam

Vietnam is selected for illustrating how JMP builds the data on drinking water access for a specific country. As can be seen from

Table 1, JMP has quite a lot of data sources for drinking water and sanitation from 1980 to 2006 with a smooth continuity of intervals. To build the trend-line for house connection in urban areas, JMP started with the data from MICS 1996 and ignored the earlier data from the 1989 National Census. However, when graphing the trend-line for total urban population using improved drinking water

Table 1: Vietnam Data Sources for House Connection and Percentage of Total Population Using Improved Drinking Water Source in Urban Areas

Source	Code	Year	House Connection		Total	
			Used for estimates	*Not used for estimates*	*Used for estimates*	*Not used for estimates*
The International Drinking Water Supply and Sanitation Decade, WHO 1984	WHO80	1980				
The International Drinking Water Supply and Sanitation Decade, WHO 1986	WHO83	1983				
The International Drinking Water Supply and Sanitation Decade, WHO 1987	WHO85	1985				70.0
The International Drinking Water Supply and Sanitation Decade, WHO 1990	WHO88	1988				48.0
	Census	1989		50.7	84.6	
The International Drinking Water Supply and Sanitation Decade, WHO 1992	WHO90	1990				47.0
Water Supply and Sanitation Sector Monitoring Report 1993	JMP93	1991				39.0
Vietnam Living Standard Survey, 1992–1993	LSS93	1993				
Water Supply and Sanitation Sector Monitoring Report 1996, WHO/UNICEF 1996	JMP96	1994				53.0

continued overleaf

Table 1: Continued

Source	Code	Year	House Connection		Total	
			Used for estimates	Not used for estimates	Used for estimates	Not used for estimates
Analysis and Evaluation of Implementation of the Mid-Term Goals for Children, 1996–1998	GSO94	1994				
Figure officially submitted to UNICEF by the Centre for Rural Water Supply and Sanitation of the Ministry of Agriculture and Rural Development after Consultation with the GSO	GSO95	1995				
Analysis and Evaluation of Implementation of the Mid-Term Goals for Children, 1996–1998. General Statistics Office and Vietnam Committee for Protection and Care of Children	MICS96	1996	43.7		91.1	
Vietnam Demographic and Health Survey 1997	DHS97	1997		66.2	94.1	
Form 4, Census 1998	CEN99	1999	43.7		91.3	
Global Water Supply and Sanitation Assessment 2000. Water Supply and Sanitation Sector Questionnaire — 1999. (Form 6 sent to WHO)	JMP99	1999		85		95.0
Analysis of results of Multiple Indicator Cluster Survey II, General Statistic Office, 2000	MICS00	2000	50.4		95.4	
Vietnam Demographic and Health Survey 2002	DHS02	2002		74	96.9	
World Health Survey, WHO 2003	WHS03	2003	49.3		96.2	
	DHS05	2005	61.1		96.6	
	MICS06	2006	58.6		97.1	

Source: JMP website (<www.wssinfo.org>) [Accessed on April 5, 2010].

sources, JMP used the data from the 1989 National Census as a starting point and then jumped to MICS data in 1996, ignoring data from WHO for 1990 and JMP for 1991 and 1994. Two immediate questions arise in this context. First, why did JMP not use other available data for 1990, 1991 and 1994 to have better estimates of the trend of population using improved drinking water sources over-time? Second, what could be the reason for JMP using data from the National Census of 1989 for total population while disregarding data for house connections? It is true that there must have been important and pressing reasons for doing so, but it is not explained on the JMP data site.

A similar problem arises in regard to DHS97 and DHS02. Data for house connections from DHS97 and DHS02 were ignored while total population estimates were used. Yet, DHS05 data for house connections were used. This could be justified if a proper evaluation of data from the three surveys was carried out and results posted on the JMP site for Vietnam. This was not the case, however.

Another serious problem is the design of survey questionnaires to collect the data on house connection, which is of course not the JMP's responsibility. Under the WHO category on "improved drinking water access", piped water into the user's dwelling (or house connection) and piped water to yard/plot are separated. However, in the WHS questionnaires, house connection and piped water to yard are integrated into one single category. The DHS questionnaires share the same design with the WHS's, while the MICS question-naires put piped water into dwelling and piped water into yard/plot in two separate categories. This inconsistency in the survey question-naires may distort the actual situation to an unknown extent when JMP used house connection data from WHS, DHS and MICS in its estimates. Moreover, where house connection has a dominant role (see the papers on IDWA), the data on combined categories can be misleading.

When one sees the used and unused data on urban house connections in Figure 1, it becomes obvious that one cannot figure out the reasons for using and not using specific data sets. The only obvious aspect that the figure demonstrates is that if all data are to be used together, then one does not discern a neat trend. Is it possible to get not so neat trends? The answer may be in the affirmative, especially for urban areas that just cannot provide proper housing and hence house connections to the newly arriving immigrants from the rural areas. One must not then be surprised to see ups and downs in the connection rates in the urban areas.

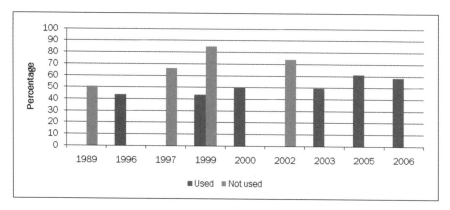

Figure 1: Vietnam Urban House Connection Data

One may wonder whether the discretions used in regard to data selection in JMP is unique for Vietnam. From Table 2, where China

Table 2: Used and Not Used Data Sources: JMP's Urban House Connection Coverage Estimates, Vietnam, China and Thailand

Vietnam		China		Thailand	
Used	Not used	Used	Not used	Used	Not used
MICS 1996	DHS 1997	CHS[2] 1989	MICS 1992	DHS 1987	JMP 1999
Census 1999	JMP 1999	CHS 1991	WHS 2003	Census 1990	
MICS 2000	DHS 2002	CHS 1997		Census 2000	
WHS 2003	Census 1989	CHS 2000		MICS 2005	
DHS 2005		CHS 2004			
MICS 2006					

Source: JMP website (<www.wssinfo.org>) [Accessed on April 5, 2010].

---

[2] CHS is the abbreviation for China Economic, Population, Nutrition and Health Survey, Household Survey

and Thailand are shown along with Vietnam, it turns out that some data sets are ignored in JMP for other countries too.

5. Summary Observation

Discrepancies from different data sources are a matter of fact because of different methodologies, definitions, sample sizes and timing. One must not assume that JMP must take all sorts of data. Yet one expects to see along with the country data pages, a clear note stating why some data are used and some are not. This is vital in assuring all concerned the international comparability of data and also the validity of the instruments used in MDG monitoring. It is time the international development community invests in data fine-tuning toward improving the monitored parameters under MDG.

References

Demographic and Health Surveys (2002). *Vietnam: DHS, 2002 — Final Report.* Available at <http://www.measuredhs.com/pubs/pub_details.cfm?ID=412> [Accessed April 5, 2010].

General Statistical Office of Vietnam (2002). *General Introduction of Vietnam Household Living Standard Survey 2002.* Available at <http://siteresources. worldbank.org/INTLSMS/Resources/3358986-1181743055198/3877319-1207074161131/General_Introduction_of_VNLSS2002.pdf> [Accessed April 5, 2010].

_____ (2004). *Household Living Standard Survey, Questionnaire on House-hold Survey.*

_____ (2004). *Living Standard Survey.* Available at <http://www.gso.gov.vn/ default_en.aspx?tabid=483&idmid=4&ItemID=4343> [Accessed April 5, 2010].

_____ (2004). *Vietnam Household Living Standard Survey 2002 and 2004.* Available at <http://siteresources.worldbank.org/INTLSMS/Resources/ 3358986-1181743055198/3877319-1207074161131/BINFO_VHLSS_02_ 04.pdf> [Accessed April 5, 2010].

_____ (2006). *Vietnam Multiple Indicator Cluster Survey.* Available at <http:// www.childinfo.org/files/MICS3_VietNam_FinalReport_2006_Eng_Vi.pdf> [Accessed April 5, 2010].

Living Standards Measurement Study (2005). *Vietnam Population and AIDS Indicator Survey.* Available at <http://www.measuredhs.com/pubs/pdf/AIS3/AIS3.pdf> [Accessed April 5, 2010].

The Joint Monitoring Program for Water Supply and Sanitation. *Vietnam: Improved water coverage estimates (1980–2008).* Available at <http://www.wssinfo.org/resources/documents.html?type=country_files> [Accessed April 5, 2010].

World Health Organization. *The World Health Survey 2002 Household Questionnaire.* Available at <https://www.who.int/healthinfo/survey/whslonghouseholdlow.pdf> [Accessed April 5, 2010].

World Health Organization. *The World Health Survey Sampling Guidelines for Participating Countries.* Available at <http://www.who.int/healthinfo/survey/whssamplingguidelines.pdf> [Accessed April 5, 2010].

World Health Organization. *World Health Survey Instruments and Related Documents.* Available at <http://www.who.int/healthinfo/survey/instruments/en/index.html> [Accessed April 5, 2010].

Xu, K., Ravndal, F., Evans, D., and Carrin, G. (2009). Assessing the reliability of household expenditure data: Results of the World Health Survey, *Health Policy*, 91, 297–305.

Annex

Brief Notes on DHS, MICS, WHS and LSMS

Demographic and Health Survey (DHS)

DHS, a global project, was initiated in 1984 and since then it provided technical assistance to more than 240 surveys in over 85 countries, advancing global understanding of health and population trends in developing countries. DHS received worldwide reputation for collecting and providing accurate, nationally representative data on fertility, family planning, maternal and child health, gender, HIV/AIDS, malaria, and nutrition.

DHS project is funded by the U.S Agency for International Development (USAID). Contributions from other donors, as well as funds from participating countries, also support DHS.

Vietnam conducted three Demographic and Health Surveys in 1997, 2002 and 2005. All three surveys were undertaken by the General Statistical Office of Vietnam (GSO) on behalf of the Population and Family Health Project of the Committee of Population, Family and Children. The Demographic and Health Surveys division of ORC Macro in Calverton, Maryland, United States, provided technical assistance to the projects through several visits and through emails. The DHS 2002 will be taken as an example to describe how the survey was done in Vietnam.

Sample Design

The sample for the DHS 2002 was based on the sample used in DHS 1997. DHS 1997, in turn, was based on the Multi Round

Demographic Survey (MRDS) which consisted of 1,590 sample areas (known as enumeration areas — EAs) spread throughout the 53 provinces/cities of Vietnam, with 30 EAs in each province. The total number of households covered in MRDS was 243,000. For the DHS 1997, a sub-sample of 205 EAs was selected with 26 households in each urban EA and 39 in each rural EA. The 2002 sample was designed to produce about 7,000 completed household interviews and 5,600 completed interviews with ever-married women from 15–49 years of age.

Questionnaire Content

Three types of questionnaires were used in the surveys: the Household Questionnaire, the Individual Woman's Questionnaire and the Community/Health Facility Questionnaire.

The Household Questionnaire was used to enumerate all usual members and visitors in selected households and to collect information on age, sex, education, marital status, and relationship to the head of the household. The main purpose of the Household Questionnaire was to identify persons who were eligible for individual interview (i.e. ever-married women aged 15–49). In addition, the Household Questionnaire collected information on characteristics of the household such as water source, type of toilet facilities, material used for floor and roof, and ownership of various durable goods. This questionnaire is the source of information regarding house connection and total population using improved water sources.

The Household Questionnaire consists of 23 questions, and questions 16 and 17 are on drinking water:

Question 16: What is the main source of drinking water for members of your household?

Possible choices of answer: Piped into residence/plot; piped to public tap; well in residence/plot; spring; river/stream; pond/lake; dam; rainwater; tanker truck; bottled water; other.

Question 17: How long does it take you to go there, get water, and come back? .... minutes.

Survey Response Rate

The following table is extracted from the Vietnam DHS 2002 report and shows us quite a high response rate.

Table 1.2 Sample results

Number of households, number of eligible women, and response rates, Vietnam 2002

Result	Residence		
	Urban	Rural	Total
Household interviews			
Households selected	1,690	5,460	7,150
Households occupied	1,664	5,392	7,056
Households interviewed	1,660	5,388	7,048
Household response rate	**99.8**	**99.9**	**99.9**
Individual interviews			
Number of eligible women	1,316	4,390	5,706
Number of women interviewed	1,300	4,365	5,665
Individual response rate	**98.8**	**99.4**	**99.3**

Multiple Indicator Cluster Survey (MICS)

MICS is an international household survey initiated by UNICEF to assist countries in collecting and analyzing data in order to fill data gaps for monitoring the situation of children and women. Since the mid-1990s, MICS has enabled many countries to produce statistically sound and internationally comparable estimates on a range of indicators in the areas of health, education, child protection and HIV/AIDS.

Vietnam conducted three Multiple Indicator Cluster Surveys in 1996, 2000 and 2006. Similar to DHS, all MICS were undertaken by GSO with the co-operation of the Population and Family Health Project of the Committee of Population, Family and Children. MICS 2006 was specifically a sample survey representing the whole country, urban and rural areas and eight geographical regions of Vietnam. The main purpose of the survey was to provide up-to-date information for assessing the situation of children and women in Vietnam, monitoring MDG goals, the National Plan of Action for Children 2001–10, and strengthening technical expertise in the survey design and implementation as well as data analysis.

Sample Design

Vietnam MICS 2006 sample was a 2-stage probability stratified and clustered sample. The first stage sample consisted of 250 sampled areas and the second stage consisted of 8,355 households selected from 8 regions and all 64 provinces/cities within the country. On average, there were 33 households in each sample area. The details of the selection of sample areas are as follows.

		Thành thị/ Urban	Nông thôn/ Rural	Cộng/ Total
	Tổng số/ Total	57	193	250
1.	Đồng bằng sông Hồng/ Red River Delta	7	25	32
2.	Đông Bắc/ North East	5	26	31
3.	Tây Bắc/ North West	4	26	30
4.	Bắc Trung bộ/ North Central Coast	4	26	30
5.	Duyên hải nam Trung Bộ/ South Central Coast	8	23	31
6.	Tây Nguyên/ Central Highlands	6	25	31
7.	Đông Nam Bộ/ South East	15	17	32
8.	Đồng bằng sông Cửu Long/ Mekong River Delta	8	25	33

Source: Extracted from The Final Report for Vietnam MICS 2006.

Questionnaire Content

There were three sets of questionnaires in MICS surveys: a household questionnaire which was used to collect information on all de jure household members, the household, and the dwelling; a women's questionnaire administered in each household to all women aged 15–49 years; and a questionnaire regarding children under 5 years of age, administered to mothers or caretakers of all children under 5 living in the household. There were several sections in each set of the questionnaire, and each section consisted of numerous questions. The example of the questions for water and sanitation is given below. It is immediately evident that the questions are quite comprehensive.

PART 1, SECTION E: WATER AND SANITATION		WS
WS1. What is the main source of drinking water of your household?		
Private piped water into house .... 11 ⇨WS5	Rain water ....51 ⎫	
Private piped water into house⬜s yard ⏋12	Water from tanker-truck ....61 ⎮	
Public piped water stand ....13 ⎰	Water from water carrying carts, ...71 ⎮ ⇨WS3	
Tube well...21	Surface water from rivers, ponds, ⎮	
Protected dug well...........................31 ⎱⇨WS3	lakes, canals ....81 ⎮	
Unprotected dug well.......................32	Bottled water ....91 ⎮ ⇨WS2	
Protected spring water.....................41	Other (specify)................................. ⎮	
Unprotected spring water ....42 ⎭	....96 ⎭ ⇨WS3	
WS2. What is the main source of water for other living purpose?		
Private piped water into house ....11⇨WS7	Rain water.......................................51	
Private piped water into house⬜s yard	Water from tanker-truck 61	
....12⇨WS7	Water from water carrying carts 71	
Public piped water stand 13	Surface water from rivers, ponds,	
Tube well..21	lakes, canals 81	
Protected dug well............................31	Other (specify)..................................96	
Unprotected dug well.......................32		
Protected spring water 41		
Unprotected spring water 42		
WS3. How many minutes does it take to go to this water source, wait, get water and return home?		
(record with 3 digits)		
Water on premises............995 ⇨WS5		
Don⬜t Know.........................998		
WS4. Who usually go to this water source to get the water?		
Probe:		
Is this person under 15 years old? Is boy or girl?		
Adult female 1		
Adult male 2		
Girl (under 15) 3		
Boy (under 15).....................................4		
Don⬜t know..8		
WS5. Do you have any treatment of the water before drinking (including boiling water)?		
Yes..1		
No..2 ⇨WS7		
Don⬜t know ....8⇨WS7		
WS6. Which treatment method have you used?		
Any other method?		
Circle all methods that have been used for water treatment		
Boil ....A	Put under sunshine ....E	
Use chemicals: chlorine, alum ....B	Let the water stand and settle ....F	
Use filtering cloth C	Other (specify).....................................X	
Use water filter system (sand, ceramic,	Don⬜t know ....Z	
coal⬜)...D		

Source: Extracted from The Final Report for Vietnam MICS 2006.

Survey Response Rate

Among 8,356 households selected for the sample, 8,355 were successfully interviewed, making a household response rate of almost 100%. In the interviewed households, 10,063 women (aged 15–49) were identified. Of these, 9,473 were successfully interviewed, yielding a response rate of 94.1%. In addition, 2,707 children under age of five were listed in the household questionnaire. Questionnaires were completed for 2,680 of these children, which corresponded to a response rate of 99%. Response rates were similar across regions and areas.

World Health Survey (WHS)

WHS was developed by the World Health Organization to compile comprehensive baseline information on the health of a population and on the outcomes associated with the investment in health systems; baseline evidence on the way health systems are currently functioning; and, ability to monitor inputs, functions, and outcomes. The objectives of WHS are to:

* Develop a means of providing low-cost, valid, reliable and comparable information.
* Build the evidence base to monitor whether health systems are achieving the desired goals.
* Provide policy-makers with the evidence they need to adjust their policies, strategies and programs as necessary.

The survey questionnaire contains 143 questions/items in the long form and 78 questions in the short form. Seventy-one countries implemented various forms of the WHS-household face-to-face surveys, computer assisted telephone interviews or computer assisted personal interviews in 2002. Sample sizes varied from 1,000 to 10,000.

There are three questions on drinking water in the questionnaire. The following table is extracted from the WHS 2002 individual questionnaire.

Q4042	What is the main source of drinking water for members of this household?	1. Piped water through house connection or yard			
		2. Public standpipe			
	(Show card to respondent ---- see Appendix A4.5)	3. Protected tube well or bore hole			
		4. Protected dug well or protected spring			
		5. Unprotected dug well or spring			
		6. Rainwater (into tank or cistern)			
		7. Water taken directly from pond-water or stream			
		8. Tanker-truck, vendor			
Q4043	How long does it take to get there, get water and come back?	1. Less than 5 minutes	2. Between 5 to 30 minutes	3. Between 30 to 60 minutes	4. Between 60 to 90 minutes	5. More than 90 minutes
Q4044	Are there at least 20 litres of water per person (about one bucket) available per day (for drinking, cooking, personal hygiene etc.) in the household?	1. Yes		5. No		

Living Standard Measurement Study (LSMS)

LSMS was established by the Development Economics Research Group of the World Bank to explore ways of improving the type and quality of household data collection by statistical offices in developing countries. The ultimate goal of LSMS is to foster the use of household data as a basis for policy and decision-making. LSMS has been working for some time to develop new methods to monitor progress in raising levels of living, to identify the consequences for households of past and proposed government policies, and to improve communications between survey statisticians, analysts, and policy makers.

In Vietnam, LSMS has got a slightly different name, Vietnam Household Living Standard Survey (VHLSS). The first two VHLSS were implemented in 1992–93 and 1997–98 by GSO with funding from the United Nations Development Program (UNDP) and technical assistance from the World Bank. Since 2000, VHLSS has been conducted every two years.

Questionnaire Content

VHLSS used three sets of questionnaire: short household question-naire (excluding most of consumption expenditure information), long household questionnaire (including detailed consumption ex-penditure information), and commune questionnaire.

The short household questionnaire contained 9 sections, each of which covered a separate aspect of household activity. Questions regarding drinking water were in section 8 about housing. Main sources of drinking water over the last 12 months were collected by area (urban/rural), gender of household head, region and five income quintiles. The categories for the sources of drinking water are: private tap; public standpipe; bottled water; pumped well; hand-dug well; filter spring water; other wells; rain collection; river, lake, pond, spring; and others.

Sample Design

In the urban domain, 700 sample areas were selected with a sample size of 25 per sample area. In the rural domain, 2,300 sample areas were selected and each sample area consisted of 20 households.

Links:

DHS: <http://www.measuredhs.com/pubs/pub_details.cfm?ID=581&ctry_id=56&SrchTp=ctry>

MICS: <http://www.childinfo.org/files/MICS3_VietNam_FinalReport_2006_Eng_Vi.pdf>

WHS: <http://www.who.int/healthinfo/survey/instruments/en/index.html>
<http://www.who.int/healthinfo/survey/whssamplingguidelines.pdf>

LSMS: <http://www.gso.gov.vn/default_en.aspx?tabid=483&idmid=4&ItemID=4343>

7

Concluding Notes

Seetharam Kallidaikurichi E. and Bhanoji Rao

The "concluding notes" are explicitly and implicitly suggested by the papers in this monograph. They refer to the need for data monitoring and possible and desirable fine-tuning of the MDG. At the end of this short note, two key messages are placed prominently to draw the attention of all concerned.

Measuring Access to Water

"Some goals cannot be met; others cannot even be measured.... And sometimes what is measured ... is not what counts...." observed *The Economist* in a lead article on the Millennium Development Goals in the issue dated July 7, 2007. Given that the achievement of each MDG is an international commitment for the betterment of humanity at large, it is expected that due attention is given to proper conceptualization of the indicators that reflect MDG targets and achievements. The various agencies that publish the indicators on access to water and sanitation bring out almost identical numbers, as evidenced by the near unanimity in the values of the indicators in respect of the Asian developing member countries of ADB (Table 1).

Table 1: Illustrative Comparison of Indicators from ADB, UNICEF, World Bank and UN

Comparing	No. of DMC covered	Difference if Any
(A) ADB Key Indicators, 2006; and	Urban: 42 for water and 41 for sanitation in A and 40 and 39 in B.	For India in A the data for urban sanitation is given as 54, while it is given as 59 in B.
(B) UNICEF Progress for Children, 2006	Rural: 40 and 39 in A and 40 and 38 in B	
(A) World Bank Global Monitoring Report 2007 and	For water – 29 in A and 40 in B.	No difference
(B) UN MDG Data Base	For sanitation – 27 in A and 39 in B	

The fact that most international agencies are using the same indicators does not necessarily mean that the data are of good quality and reliability. This has been amply demonstrated by the Vietnamese case study in this volume. Data evaluation exercises do not seem to have been carried out at national/international levels, even sparingly if not routinely.

That the current water access indicators are not quite reflective of what is desirable for the people's lives is brought out neatly by the weak correlations between house connections and access rates across countries. More the house connections more should be access. Empirical reality is far from that.

From the Water and Sanitation Information Website of the Joint Monitoring Program of WHO and UNICEF, for the Asia-Pacific countries, we could assemble, for one or more years between 1987 and 2003, data on the percentage of families served by house connections for water and sanitation. Juxtaposing this data to the average water and sanitation access rates for 1990 and 2004, correlation coefficients are obtained as shown in Table 2.

Table 2: Correlations between House Connections and Access Rates ('n' refers to sample size)

	Urban	*Rural*
Water [n]	0.65 (22)	0.40 (20)
Sanitation [n]	0.49 (19)	0.39 (16)

The correlations indicate in general the relatively weak link between house connections and overall access rates. Two possible interpretations can be suggested. One is that it does not matter if a house connection is available or not for defining access. This may not be an acceptable proposition if one were to ensure the same idea and basket of goods and services to constitute development for one and all.

Another and relatively more important interpretation of the correlations is that access estimates, since they are not fully reflective of the availability of house connections, must be treated as aggregations of diverse water/sanitation facilities, which may or may not be internally consistent and internationally comparable.

House Connection Implies Having a House

House connections for water and sanitation go beyond providing access; they indicate a minimum level of decent living. In order to provide for such a living standard, it is useful to specifically incorporate within the MDG Goal 7, Target 11, the target of "housing the homeless", homeless being inclusive of those without proper housing, as part of the goal to provide sustainable and safe/good water and sanitation on an economically viable basis, with cross subsidization as needed.

Key Messages

1. The international community has the obligation to ensure the availability, reliability and international comparability of the

indicators of access to water and sanitation. It may be best to constitute a panel of experts to look into the issue and come up with widely acceptable recommendations.

2.   Housing for Each and Every Family [HEEF] ought to be the distinct goal for the world at large to be achieved by a not too distant date.

About the Authors

Seetharam Kallidaikurichi E.

Visiting Professor and Founding Director, Institute of Water Policy, Lee Kuan Yew School of Public Policy and founding Director of the NUS Global Asia Institute.

Seetharam Kallidaikurichi E. is an internationally recognized expert with 20 years of professional experience on development cooperation, infrastructure, integrated planning for economic growth, participatory social development, diplomacy, and Human Values. He is seconded from the Asian Development Bank (ADB) where he was Principal Water Supply and Sanitation Specialist, and a focal point guiding ADB's operations in water supply, sanitation, and wastewater management in line with ADB's Water Policy.

Bhanoji Rao

Visiting Professor, Institute of Water Policy, Lee Kuan Yew School of Public Policy

Bhanoji Rao holds a Doctorate in Economics from the University of Singapore. His teaching and research career spawning over 4 decades included positions at the University of Singapore, the National University of Singapore and the World Bank. He published 17 books, 40 chapters in books, 62 papers in refereed journals and 14 monographs. Presently he is also Visiting Faculty, Sri Sathya Sai University, Prasanthi Nilayam, and ASCI, Hyderabad.

Fan Mingxuan

Research Associate, Institute of Water Policy, Lee Kuan Yew School of Public Policy

Fan Mingxuan received a Master in Public Policy from Lee Kuan Yew School of Public Policy, National University of Singapore. Her current research focuses on water and related data and statistics, water pricing, and water supply and sanitation with special reference to China. Prior to joining the Institute of Water Policy, she worked for the ADB's Greater Mekong Sub-region Environment Operations Center in Bangkok.

Ngo Quang Vinh

Research Associate, Institute of Water Policy, Lee Kuan Yew School of Public Policy

Ngo Quang Vinh received a research scholarship in 2007 from the National University of Singapore for his Master of Social Science degree in economics. He had close to ten years of work experience as a lecturer of International Business and Economics at Foreign Trade University in Hanoi, Vietnam from where he received his Master and Bachelor degrees in Economics.

Index